ALGEBRA

algebra

P. Abbott

Revised by Michael Wardle

TEACH YOURSELF BOOKS

For UK order queries: please contact Bookpoint Ltd, 39 Milton Park, Abingdon, Oxon OX14 4TD. Telephone: (44) 01235 400414, Fax: (44) 01235 400454. Lines are open from 9.00 - 6.00, Monday to Saturday, with a 24 hour message answering service. Email address: orders@bookpoint.co.uk

For U.S.A. & Canada order queries: please contact NTC/Contemporary Publishing, 4255 West Touhy Avenue, Lincolnwood, Illinois 60646-1975, U.S.A.. Telephone: (847) 679 5500, Fax (847) 679 2494.

Long-renowned as the authoritative source for self-guided learning – with more than 30 million copies sold worldwide – the *Teach Yourself* series includes over 200 titles in the fields of languages, crafts, hobbies, sports, and other leisure activities.

British Library Cataloguing in Publication Data
Abbott, P.
  Algebra. – 4Rev.ed
  I. Title II. Neill, Hugh
  512

Library of Congress Catalog Card Number: On file.

First published in UK 1996 by Hodder Headline Plc, 338 Euston Road, London, NW1 3BH.

First published in US 1996 by NTC/Contemporary Publishing, 4255 West Touhy Avenue, Lincolnwood (Chicago), Illinois 60646-1975 U.S.A.

The 'Teach Yourself' name and logo are registered trade marks of Hodder & Stoughton Ltd in the UK.

Typeset by Wearset, Boldon, Tyne & Wear.
Printed in Great Britain for Hodder & Stoughton Educational, a division of Hodder Headline Plc, 338 Euston Road, London NW1 3BH by Cox & Wyman Ltd, Reading, Berkshire.

Impression number   14 13 12 11 10 9 8 7 6 5
Year   2003 2002 2001 2000 1999 1998

Contents

Introduction

Algebra is such a wide and comprehensive subject that this book cannot be regarded as anything more than an elementary introduction to it. It will enable you to learn something of the principles and foundations of the subject, thus enabling you to proceed to the study of more detailed and advanced books. It also provides the knowledge of algebra required for applications of mathematics to engineering, etc. Consequently some of those elementary sections of the subject which are of little use for these purposes have not been included.

The exercises are progressive and designed both to enable you to test your knowledge of the work you have studied and also to provide material for your training in that power of manipulation which is so essential. They contain few of the more complicated or academic problems which are beyond the practical requirements of the ordinary student.

An appendix contains, without exercises, a very brief summary of the meaning of permutations and combinations, the binomial theorem, and the nature of the roots of a quadratic equation. Also included are those formulae which you may require when beginning work on the calculus or other branches of mathematics.

While the fundamental laws of algebra have not been entirely overlooked, rigid proofs of them have been omitted, owing to exigencies of space. It is hoped, however, that the logical basis of the subject has not been seriously impaired by the omissions.

Some emphasis has been placed on the graphical aspects of parts of the subject, since experience has shown that they prove stimulating and provide revealing help to the student.

No previous mathematical knowledge is required for this work, beyond that of arithmetic. References have occasionally been made to theorems in geometry or trigonometry for the benefit of those students who have some knowledge of them.

The authors would like to thank Mr Hugh Neill for his work on the new edition. This includes redrawing the illustrations and thoroughly updating and modernising the text. Thanks too to Mr C. E. Kerridge for permission to use examples from *National Certificate Mathematics* Vol I and Mr H. Marshall for the use of examples from Vol II.

I

The Meaning of Algebra

I ALGEBRA AND ARITHMETIC

Algebra, like arithmetic, deals with numbers. The two subjects have much in common; indeed, algebra has been called 'generalised arithmetic', though this is a very incomplete description of it. It would perhaps be more correct to say that algebra is an extension of arithmetic.

Both subjects employ the fundamental operations of addition, subtraction, multiplication and division of numbers, subject to the same laws. In each, the same symbols, $+$, $-$, $\times$, $\div$, are used to indicate these operations, but in algebra, as new processes are developed, new symbols are invented to assist the operations. Terms such as fractions, ratio, proportion, square root, etc., have the same meaning in both subjects, and the same rules govern their use.

In arithmetic you use **numbers** whose values are known; you operate with these and obtain definite numerical results. Whereas in algebra, while you may use definite numbers on occasions, you are, in the main, concerned with **general expressions** and **general results**, in which letters or other symbols represent numbers not named or specified.

This may seem vague to a beginner, but the following example may serve to show what is meant.

2 A FORMULA

In arithmetic you learn that to find the area of the floor of a room, rectangular in shape, you 'multiply the length of the room by the breadth'. This might be expressed in the form:

The area of a rectangle in square metres is equal to the length in metres multiplied by the breadth in metres.

This rule is shortened in algebra by employing letters as symbols, to represent the quantities. Thus:

Let the letter l represent 'the length in metres'.

Let the letter b represent 'the breadth in metres'.

Let the letter A represent 'the area in square metres'.

With these symbols you can now write the above rule in the form:

$$A = l \times b$$

This shortened form shows the rule for finding the area of any rectangle; it is a general rule, and is called a **formula**. The full description of it would be

'The formula for finding the area of a rectangle.'

Notice that in the formula above there is no mention of units. This is because it is true whatever units are employed. It is necessary, however, that you use the same kind of unit throughout. If l and b are measured in mm, A will be the area in square mm; if they are measured in metres, A will be in square metres.

When using the formula for any specific case it is important to state clearly what unit is employed for l and b. The unit for A will then follow.

Algebra has been called a kind of shorthand, and the above example in which a sentence has been reduced to

$$A = l \times b$$

illustrates the reason for the description, but as you make progress you will discover that it is an incomplete description. From such beginnings, the subject develops into the most powerful instrument employed in mathematics.

The Greeks produced some of the greatest mathematicians in history, but their work was mainly accomplished in geometry. They made very little progress in algebra, because they did not have appropriate symbols. They did not have separate symbols for numbers; these symbols were later introduced by the Arabs, and are still in use today. The Greeks used instead the letters of the alphabet to represent numbers. The Romans were similarly restricted, and neither nation employed the decimal system of notation. You will realise something of the great difference suitable symbols make in mathematics if, as an example, you write down the number of the year in Roman numerals and then try to multiply it, say by 18, also expressed in Roman fashion.

Progress in manipulation and the general development of the subject requires symbols, which must be suitable for their purpose. The history of mathematics reveals that many of the symbols which are now so familiar have been reached by a slow evolutionary process, often lasting through centuries.

When letters are used to represent numbers you can make any suitable choice. In the above formula, although A, l and b were employed, any other letters could be used if they were more suitable. By common usage, however, certain letters are usually selected for specific purposes, and the same symbols are used invariably to denote certain numbers. This will become clear to you as you progress.

But whatever letters you use to solve problems, you must observe the following rules:

(1) you must state clearly what each symbol represents;
(2) if measurements of any kind are involved, you must define clearly the units you employ.

3 TRANSFORMATION OF A FORMULA

You may be inclined to wonder why such an elementary rule as the area of a rectangle should be expressed as a formula by the employment of algebraic symbols. But this is an example selected for its simplicity to illustrate the meaning of an algebraic formula. Even in this case it is possible in some measure to illustrate the flexibility and adaptability of a formula as compared with a statement in words.

For example, suppose that l and b were measured in cm, then $A = l \times b$ square centimetres.

If you want to express this result in square metres you could do it by writing:

$$A = \frac{l \times b}{10\,000} \text{ square metres}$$

Suppose that the rectangle is a room, l and b being measured in cm, and that the room is covered with carpet costing a pence per square metre. If the cost of the whole carpet is represented by C pence, then:

since $$A = \frac{l \times b}{10\,000} \text{ square metres}$$

then $$C = £\frac{l \times b}{10\,000} \times a \text{ pence}$$

If this were to be expressed in pounds, then:

$$C = £\frac{l \times b}{10\,000} \times \frac{a}{100}.$$

Thus a formula may be used as a foundation for other formulae to express modifications of the original.

There is a convenient notation which is employed throughout mathematics and it is more fully discussed in Section 18. Briefly, when a number is multiplied by itself, this is called squaring the number: thus, $a \times a$ is a squared and is written a^2. Similarly, 1 square metre, the unit of area which is $1\text{ m} \times 1\text{ m}$, is written as 1 m^2. 28 square metres would be written 28 m^2. Again, m^3 is $\text{m} \times \text{m} \times \text{m}$ and represents cubic metres.

4 EXAMPLES OF USING FORMULAE

Foreign currency

When travelling abroad you usually have to change the money of your own country into that of the country that you are visiting. A formula is used to convert one currency into the other. In order to do this you need to know the current exchange rate. For example, in October 1995, the number of German marks you might get in exchange for £1 was 2.25. This can be shown as the formula: *Number of marks is number of pounds times* 2.25, or $M = 2.25 \times S$, where M is used to represent the number of marks and S is used to represent the number of pounds sterling. To find the number of marks you would get for £50 you replace the letter S in the formula by 50. So the number of marks you would get is 2.25×50 i.e. 112.5 marks. If you wanted to change marks into pounds then you would have to rewrite the formula as $S = M/2.25$. To find the number of pounds you have to divide the number of marks by 2.25.

Income tax

Income tax is based on your income after certain allowances have been deducted. In 1995 a single person was allowed to earn at least £3,445 before paying any tax. To work out the tax due on an income of £11,425, you first subtract the allowance of £3,445 from the income of £11,425, and then find 25% of the figure remaining, i.e. 25% of (11 425 − 3445). In this case the tax due is 25% of £7,980 which is £1,195. This can be shown as the formula: *Tax due is 25% of (income minus allowances)* or $T = 25\%$ of $(I − A)$, where T is used to represent the tax due, I is the total income and A are the deductable allowances.

Car hire

Often when hiring a car you are charged a standard amount plus so much for each mile. For example the charge might be £24 plus 30p for each mile that you use the car. In this the charge for a total journey of 80 miles would be £24 + 80 × 30p, i.e. £48. This can be shown as the formula: *Total cost is standard charge plus number of miles times* 30p or $C = S + M × 0.30$, where C is the total charge, assumed to be in pounds, S is the standard charge and M is the number of miles travelled. Note that you must write the 30p as 0.30 in the formula so that the units of each of the terms are the same.

5 AN ILLUSTRATION FROM NUMBERS

The following example illustrates the use of letters to represent generalisations in number. You know that:

(1) if any integer – i.e., a whole number – is multiplied by 2, the result is always an even number;
(2) if any even number is increased by unity, the result is an odd number.

These two statements can be combined in one as follows.

If any integer is multiplied by two and the product increased by unity, the result is an odd number.

This is a generalisation about an odd number, expressed in words. This can be expressed by means of algebraical symbols as follows.

(1) Let n be any integer.

(2) Then $2 \times n$ is always an even number
(3) Therefore $2 \times n + 1$ is always an odd number.

Step (3) shows an algebraic expression which you can use to represent any odd number.

The brevity and lucidity of this expression as compared with the full description of an odd number above will be apparent. But its value goes beyond this. You can manipulate this algebraic form, you can operate with it, and in this way, you can use it in the solution of problems.

First note, however, that when expressing the product of two or more numbers represented by letters or a numeral and letters, you can leave out the multiplication sign. Thus $2 \times n$ can be written as $2n$, and $2 \times n + 1$ as $2n + 1$. This cannot be done with two numerals, such as 25, because under the decimal system the figure 2 has a place value. Multiplication may also be shown by a dot, thus $2.n$.

If any odd number can be represented by $2n + 1$, then, since when any odd or even number is increased by 2 the result is the next odd or even number, the next odd number greater than $2n + 1$ is $2n + 1 + 2$ or $2n + 3$.

Similarly $2n + 5$ is the next odd number above $2n + 3$.

Consequently the expressions

$$2n + 1, 2n + 3, 2n + 5, 2n + 7 \ldots \text{etc.},$$

represent a sequence, or a succession of consecutive increasing odd numbers.

Similarly, if you subtract 1 from an even number, you obtain an odd number.

$2n - 1, 2n - 3, 2n - 5 \ldots$, represents a sequence of decreasing odd numbers.

Note: The succession of 'dots' after the sets of odd and even numbers indicates that you could write down more such numbers if it were necessary.

6 SUBSTITUTION

In the algebraic representation of a set of odd numbers – namely:

$$2n + 1, 2n + 3, 2n + 5 \ldots$$

since n represents any integer, you could, by assigning some particular value to n, obtain the corresponding odd number.

Thus if $\qquad n = 50$
then $\qquad 2n + 1 = 2 \times 50 + 1$
$\qquad\qquad\qquad = 100 + 1$
$\qquad\qquad\qquad = 101.$
Similarly, $\qquad 2n + 3 = 2 \times 50 + 3$
$\qquad\qquad\qquad = 100 + 3$
$\qquad\qquad\qquad = 103.$

Consequently the sequence of increasing odd numbers corresponding to this particular value of n is:

$$101, 103, 105, 107 \ldots$$

Similarly, the decreasing odd numbers when $n = 50$ can be found by substituting this value of n in the sequence:

$$2n - 1, 2n - 3, 2n - 5 \ldots$$

Then you get the corresponding arithmetical values:

$$99, 97, 95 \ldots$$

7 EXAMPLES OF GENERALISING PATTERNS

There are many situations which you can describe by using an algebraic expression or formula. Here are three examples which show how you can describe a typical pattern in a sequence by using letters.

Look at the pattern of dots in each of the shapes in Fig. 1 below.

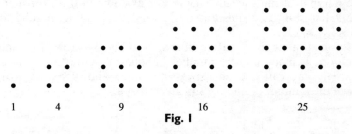

1　　　　4　　　　9　　　　　16　　　　　25
Fig. I

In the second pattern there are $2 \times 2 (= 4)$ dots, in the third pattern there are $3 \times 3 (= 9)$ dots, in the fourth pattern there are $4 \times 4 (= 16)$ dots. In the next pattern there are $5 \times 5 (= 25)$ dots. If the pattern continued in the same way there would be 10×10 $(= 100)$ dots in the tenth shape. To find the number of dots in any particular pattern you find the position of the pattern in the sequence and then multiply this number by itself. The number of dots in the seventeenth pattern in the sequence would be

$17 \times 17 \, (= 289)$ dots. In a similar way in the *n*th pattern there would be $n \times n$ (or n^2) dots.

Look at the pattern of shapes in Fig. 2, below. Each one is made up of a different number of lines. The first has 3 lines, the second has 5 lines, the third has 7 lines, the next again has two more lines and so on. These are the odd numbers. The number of lines in each shape is two more than the number in the previous shape.

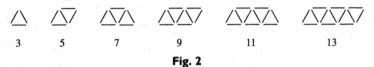

Fig. 2

You can relate the number of lines in any given shape to the position of the shape in the sequence. The fourth shape has $2 \times 4 + 1 \, (= 9)$ lines, the fifth shape has $2 \times 5 + 1 \, (= 11)$ lines, the sixth shape has $2 \times 6 + 1 (= 13)$ lines, whilst the tenth shape would have $2 \times 10 + 1 \, (= 21)$ lines.

To find the number of lines in any particular shape you find the position of the shape in the sequence, double this number and then add one.

You can use letters to describe the number of lines in a particular shape. If you use the letter *n* to describe the position of the *n*th shape in the sequence, then you can write the number of lines as $2 \times n + 1$, or $2n + 1$. You can use this expression to write down the number of lines in any shape. For example in the 17th shape the value of *n* to be substituted in $2n + 1$ is 17, so the number of lines is $2 \times 17 + 1 \, (= 35)$ lines.

Look at the second pattern of shapes in Fig. 3, below. This time each one is made up of a number of squares. As in the previous example you can find an expression which allows you to write down the number of lines in any given shape.

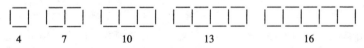

Fig. 3

The number of lines increases by three for each new shape. You can relate the number of lines in any given shape to its position in the sequence. The first shape has $1 + 3 \, (= 4)$ lines, the second shape has $1 + 3 + 3 \, (= 7)$ lines, the third shape has $1 + 3 + 3 + 3 \, (= 10)$ lines, whilst the tenth shape would have $1 + 10 \times 3 \, (= 31)$ lines.

To find the number of lines in any particular shape you find the number of squares in the shape, multiply this number by 3 and then add 1.

You can use letters to describe the number of lines in a particular shape. If you use the letter n to describe the number of squares in the nth shape in the sequence, then you can write the number of lines as $3 \times n + 1$, or $3n + 1$. You can then use this expression to write down the number of lines in any shape. For example in the 17th shape the value of n to be substituted in $3n + 1$ is 17, so the number of lines is $3 \times 17 + 1$ ($= 52$) lines.

There are various ways of looking at the number of lines in the pattern of squares above. Each of these can lead to a different way of describing the typical shape. For example in each shape the number of horizontal lines is double the number of squares in the shape and the number of vertical lines is one more than the number of squares. If you use the letter n to represent the number of squares in the shape then the total number of lines is $2n + (n + 1)$ which you can also write as $3n + 1$, as before. Another way might be to take the number of lines in each of the n squares and then to subtract the number of vertical lines where two adjacent squares touch, since these have been counted twice. The number of lines in n squares is $4n$. The number of lines counted twice is one less than the number of squares i.e. $n - 1$, so the total number of lines in the n squares is $4n - (n - 1)$ which can also be rewritten as $3n + 1$, as before.

Using letters to describe the general case, in situations like these, is a very powerful way of summarising what might otherwise be an extremely lengthy process. Algebraic expressions and formulae are the mathematician's shorthand which in many cases allow information to be both concise and extremely precise at one and the same time.

8 LETTERS REPRESENT NUMBERS, NOT QUANTITIES

Such things as length, weight, cost, are called quantities. In short, anything which can be measured is called a quantity.

Letters are not used to represent these quantities. Thus in the formula which we stated above – namely:

$$A = l \times b$$

l stands for the **number** of cm or m, as the case may be, and to obtain it you must first determine the unit to measure the quantity, and then use a letter to represent the number of units.

It is important, therefore, that when you use letters in algebraical expressions, you clearly state what each letter represents.

Thus you should say:

Let *l* represent the number of cm in the length or, more briefly, let the length be *l* cm.

Similarly let *n* represent the number of men, and let *c* represent the cost in pence.

Sometimes, for brevity, and somewhat loosely, people speak, for example, of the area of a rectangle as *A*, when they should say *A* m^2 or *A* units of area.

You can use the sign = loosely in abbreviating the statements above. Thus you could write:

Let *l* = the length in metres,
or let *c* = the cost in pence.

The sign = means 'equals' or 'is equal to'. It should connect two expressions which are equal in magnitude, but is often used loosely as above to express equality.

9 EXAMPLES OF ALGEBRAIC FORMS

Here are a few examples of what are called algebraic forms – i.e., the expression in algebraic symbols and signs of operation of statements about quantities.

■ *Example 1:* Express in algebraic form the number of pence in *x* pounds added to *y* pence.

To express pounds in pence you multiply by 100.

x pounds = $100 \times x$ pence,
and the total number of pence is $100x + y$.

■ *Example 2:* A car travels for *t* hours at *v* km/h. How far does it go? How far will it go in 20 min?

The car goes *v* km in 1 hour and so 2*v* km in 2 hours, 3*v* km in 3 hours and so on; in *t* h it will go $t \times v$ km.

20 min is $\frac{1}{3}$ hour and so the car travels $\frac{1}{3} \times v$ km in 20 min.

In $\frac{1}{3}$ hour the car travels $\dfrac{v}{3}$ kilometres.

■ *Example 3:* There are two numbers; the first is multiplied by 3 and 5 is added to the product. This sum is divided by 4 times the second number. Express the result in algebraic form.

Begin by choosing letters to represent the unknown numbers.

Let x represent the first number.

Let y represent the second number.

Then three times x increased by 5 is represented by $3x + 5$, and four times the second number is $4y$.

The division of $3x + 5$ by $4y$ is written as $\dfrac{3x + 5}{4y}$.

EXERCISE 1

1 Write down expressions for:
    (1) the number of pence in £x;
    (2) the number of pounds in n pence.

2 If £a is divided among n boys, how many pence will each boy get?

3 If n people subscribe £a each and m other people subscribe b pence each, how many pence are subscribed in all?

4 Write down the number of
    (1) metres in a kilometres;
    (2) pilots in x a jumbo jet;
    (3) tonnes in y kilogrammes.

5 The sum of two numbers is 28. If one number is n, what is the other?

6 The difference between two numbers is x; if one of them is 50, what is the other?

7 The product of two numbers is a and one of them is x; what is the other?

8 If the average length of a woman's step is x cm,
    (1) how far will she walk in 100 steps;
    (2) how many steps will she take in walking 1.8 km?

9 A number (n) is multiplied by 2, and 5 is added to the sum. Write down an expression for the result.

10 If x is an odd number write down expressions for
    (1) the next odd numbers above and below it;
    (2) the next even numbers above and below it.

11 A farmer buys sheep; x of them cost a pence each and y of them cost b pence each. What was the total cost in pounds?

12 A number is represented by x; double it, add 5 to the result, and then divide the whole by $6y$. Write down an expression for the result.

13 What number must be subtracted from a to give b?

14 What number divided by x gives y as a quotient?

15 What is the number which exceeds b by a?

16 The numerator of a fraction is x increased by 2. The denominator is y diminished by 5. Write down the fraction.

17 A car travels for m hours at v km/h. It then travels n hours at u km/h. How far does it travel in all?

18 What is the total number of pence in

$$£a + b \text{ pence?}$$

19 A train travels at v km/h. How far does it go in x hours and how long does it take to go y km?

20 Two numbers, less than 10, are chosen. Add them; multiply this sum by 2; add 4; multiply by 3; add 4 times one of the original numbers; take away 12; take away 5 times the other original number. The result is $10m + n$. What were the original numbers? Interpret this result as a party trick.

2

Elementary Operations in Algebra

10 USE OF SYMBOLS

In order that you may become familiar with the processes of algebra you need considerable practice in the use of symbols. Consequently in this and subsequent chapters you will constantly be using letters which represent numbers in a general way and without any reference to quantities such as length, cost, etc., as was done in the previous chapter.

Thus when the form $a + b$ is used, it will, in general, be using the letters, not as referring to any particular quantity, but as standing for any numbers.

11 SYMBOLS OF OPERATION

As stated in Section 1, certain symbols of operation, such as $+$, $-$, $\times$, $\div$, $\sqrt{\ }$, are common to arithmetic and algebra, since they are used for operations which are performed in both subjects. Usually, however, there is a certain difference in the way they are employed. It is evident that while such operations as

$$5 + 7, 10 - 3, 6 \times 4, 15 \div 3, \sqrt{9}$$

can be, and usually are, carried out at once with definite numerical results, expressions such as

$$a + b, a - b, a \times b, a \div b, \sqrt{a}$$

cannot be evaluated numerically while a and b represent any numbers. Until numerical values are assigned to a and b you cannot proceed further with the operation. But you can, and will, operate with the expressions themselves, without any reference to their numerical values.

In addition to the above, many other symbols of operation are used in Algebra, among them the following:

Symbol	Meaning
$=$	See Section 8
$\neq$	is not equal to
$\approx$	is approximately equal to
$>$	is greater than
$<$	is less than

12 ALGEBRAIC EXPRESSION. TERMS

A combination of letters and symbols such as $2a + b$ is an example of an algebraic expression. It may be defined as follows:

An algebraic expression is a combination of symbols which stand for numbers and for operations with them.

For brevity the term 'expression' is usually employed. When the expression contains the symbols of operation $+$ or $-$, those parts of the expression which they separate are called terms.

Thus $2a + 3b$ is an expression of two terms or a binomial.

$\dfrac{5x}{2} - \dfrac{3y}{5} + 6z$ is an expression of three terms or a trinomial.

A combination of letters such as $\dfrac{5ab}{6}$ which does not contain either of the signs $+$ or $-$ is an expression of one term, or a monomial.

13 BRACKETS

It frequently happens that an expression, or part of an expression, is to be operated as a whole. For example, suppose that you wish to write in algebraic symbols 'Twice the sum of a and b'.

Evidently the arrangement $2 \times a + b$ does not make it clear whether the 2 is to multiply a only or the sum of a and b.

Consequently 'brackets' are used to enclose the part which is to be operated on as a whole – namely, $a + b$.

$\therefore$ you write $2(a + b)$.

In this arrangement the multiplication sign is omitted between the 2 and the bracket.

The brackets have the effect of indicating the order in which you carry out the operations. Thus:

$$2(a + b) - c$$

means that you find the sum of a and b, multiply this by 2 and then subtract c.

Similarly $(a + b) \times (c + d)$ or $(a + b)(c + d)$ means that you find the sum of a and b, and also of c and d and then multiply the two results.

This will be considered further at a later stage.

14 COEFFICIENT

The expression $3a$ denotes a multiple of a and the number 3, which indicates the multiple, is called the coefficient of a.

The coefficient may be a definite number like 3, called a numerical coefficient, or it may be a letter representing a number.

Thus in the expression ax, you may regard a as the coefficient of x, but in some problems when you are thinking of multiples of a, x would be a coefficient of a; in such a case you would usually write the expression as xa.

In general, if an expression is the product of a number of factors, any one of them can be regarded as the coefficient of the product of the others, when for any purpose we regard this product as a separate number.

Thus in $3ab$, 3   is the coefficient of ab
 $3a$ is the coefficient of b
 $3b$ is the coefficient of a.

In an expression, terms which involve the same letter, and differ only in the coefficients of this letter, are called **like terms**.

Thus in the expression

$$3a + 5b - 2a + 4b$$
 $3a$ and $2a$ are like terms
and $5b$ and $4b$ are like terms.

15 ADDITION AND SUBTRACTION OF LIKE TERMS

In arithmetic you learn that the sum of

	5 dozen and 9 dozen is 14 dozen,	
or	$(5 \times 12) + (9 \times 12) = 14 \times 12.$	(A)

Similarly	$8 \text{ score} + 7 \text{ score} = 15 \text{ score}$
or	$(8 \times 20) + (7 \times 20) = 15 \times 20.$

So for any number as for example

$$(6 \times 24) + (11 \times 24) = 17 \times 24.$$

In algebra, if you were to let *a* represent 12 in the statement (A) given at the beginning of this section, you could write

$$5a + 9a = 14a,$$

and for the other cases

$$8a + 7a = 15a$$

and $$6a + 11a = 17a.$$

These last three cases are generalised forms of the preceding examples, but it must be noted that whereas in the arithmetical forms you can proceed to calculate the actual value of the sum in each, in the algebraical forms you can proceed no further in the evaluation until a definite numerical value is assigned to *a*.

Subtraction leads to similar results, just as

$$9 \text{ dozen} - 5 \text{ dozen} = 4 \text{ dozen}.$$

so $$9a - 5a = 4a.$$

In this way you can add or subtract like terms only. It is not possible, for example, to perform any addition of two unlike terms such as $9a + 5b$.

The rule for adding together like terms will now be clear. It is 'add the coefficients'.

Thus the sum of $2x + 5x + 3x = 10x$, whatever x may be. The operation of 'finding the sum' is used to include both addition and subtraction. This is called the 'algebraic sum'.

When an expression contains more than one set of like terms, you collect them and deal with them separately.

16 WORKED EXAMPLES

■ *Example 1:* Simplify $5a + 6b + 2a - 3b$.

Collecting like terms $5a + 2a = 7a$

$6b - 3b = 3b.$

Hence the whole expression is equal to $7a + 3b$. In practice there is generally no need to write down the above steps. You can make the calculations mentally.

■ *Example 2:* Simplify $15x - 3y + 6y + 7x - 5$.

Collecting like terms and adding coefficients you get

$$22x + 3y - 5.$$

17 THE ORDER OF ADDITION

If you count a number of things the total is not affected by the order in which you count them. Thus:

6 apples + 4 apples

is the same in number as

4 apples + 6 apples.

This will be clear when you remember that 6 is the symbol for 6 units, and 4 is the symbol for 4 units.

Thus:

$$6 + 4 = (1 + 1 + 1 + 1 + 1 + 1) + (1 + 1 + 1 + 1)$$

and

$$4 + 6 = (1 + 1 + 1 + 1) + (1 + 1 + 1 + 1 + 1 + 1).$$

In each case the total number of units is the same.

Thus algebraically $6a + 4a$ is the same in value as $4a + 6a$. This is true for any algebraical sum.

Thus $6a + 5b - 3a$ can be written as $5b - 3a + 6a$ without altering the value of the expression.

Briefly the order in which numbers may be added is immaterial.

18 EVALUATION BY SUBSTITUTION

If you wish to find the numerical value of an algebraical expression for definite numerical values of the letters in it, you should first simplify the expression by adding like terms. Then substitute the numerical values for the letters.

■ *Example:* Find the value of $6x + 2y - 3x + 4y - 3$ when $x = 3$ and $y = 2$.

Simplify the expression as in Section 15:

$$6x + 2y - 3x + 4y - 3 = 3x + 6y - 3.$$

Substituting the given values:

$$\begin{aligned} 3x + 6y - 3 &= (3 \times 3) + (6 \times 2) - 3 \\ &= 9 + 12 - 3 \\ &= 18. \end{aligned}$$

Notice that brackets are introduced when it is desirable to keep terms separate for evaluation.

EXERCISE 2

1 (a) Find the value of 6 dozen + 4 dozen.
   (b) Simplify $6a + 4a$ and find its value when $a = 12$.
2 (a) Find the value of $(8 \times 73) - (3 \times 73)$.
   (b) Find the simplest form of $8b - 3b$ and find its value when $b = 73$.
3 Write down in its simplest form:

$$a + a + a + b + b + b + b$$

and find its value when $a = 5$ and $b = 8$.
4 Add together $2a$, $4a$, a, $5a$ and $7a$ and find the value of the sum when $a = 2.5$.
5 Write the following expressions in their simplest forms:

   (1) $15b + 11b$ (2) $15x - 3x + 7x$
   (3) $9a - 4a + 6a + a$ (4) $4x + 3x - 2x - x$

6 Write in their simplest forms:

   (1) $5a - 2b - 3a + 6b$ (2) $11p + 5q - 2q + p$
   (3) $a - 2 + 3b + 6 + 5a$

7 Add together:

   (1) $4a - 5b, a + 6b, 5a + b$
   (2) $b + c - 3d, c + 2b + d, d - b - c$
   (3) $5x + 2y + 3z, x - y - 2z, 2x - y + z$

8 When $a = 2, b = 3$ find the numerical values of

   (1) $3a + 2b + 1$ (2) $5a - 3b + 6$
   (3) $6a + 2b - 3a + 1$ (4) $4a - 5b - 2b + 12a$

9 Simplify the following expressions and find their values when $a = 4, b = 2, x = 3, y = 5$.

    (1) $4ab - 2ab + 6ab$
    (2) $5ax - 2ax + bx$
    (3) $6xy - 4xy + xy$
    (4) $ab + 6bx - ay + 3ab - 2bx$

10 Find the numerical value of $a + \frac{1}{2}a + \frac{1}{4}a + \frac{1}{8}a$ when $a = 2$.

11 When $x = 1, y = 2, z = 3$, find the numerical value of

$$3x + 5y - 4z + 8x - y + 5z.$$

12 When $x = 4, y = 5, z = 1$, find the values of

    (1) $3xy + 2yz - 3z - 1$ (2) $xy + yz + zx$
    (3) $xy + y + x + 1$

13 If n is an odd number, write down the next three odd numbers greater than it and find their sum.

14 Write down a sequence of four numbers of which the first is a, and each of the others is twice the one which precedes it. Find their sum.

15 Write down a sequence of 5 numbers of which the first is a, the second is greater than the first by d and each of the other three is greater by d than the one which precedes it. Find their sum.

16 There are 5 numbers, the smallest of which is expressed by $2n + 5$. Each of the others is 3 greater than the one which precedes it. Write down the numbers and find their sum.

19 MULTIPLICATION

In algebra, as in arithmetic, the multiplication of a number of factors may be performed in any order, or, more precisely:

The product of a number of factors is independent of the order in which they are multiplied.

Thus 3×4 is equal in value to 4×3
             $6 \times 3 \times 5$ is equal in value to $3 \times 5 \times 6$

and generally $a \times b$ is equal in value to $b \times a$

and $a \times b \times c$ is equal in value to $c \times b \times a$.

Consequently if you require to multiply say, $2a$ by 5 you can write the product in the following form:

$$2 \times a \times 5 = 2 \times 5 \times a$$
$$= 10a$$

and

$$3a \times 2b = 3 \times 2 \times a \times b$$
$$= 6ab.$$

In this last example, a and b being unlike letters, you cannot proceed further with the multiplication.

You should note, however, that although 4×3 is equal in value to 3×4, the two products do not necessarily mean the same thing when they refer to quantities.

If, for example, 12 soldiers were to 'form fours', they would be arranged as shown in Fig. 4(a), the arrow showing the direction in which they are facing. But if the same 12 soldiers were to 'form threes' they would be arranged as in Fig. 4(b).

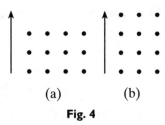

(a)   (b)

Fig. 4

Thus 3 rows of 4 soldiers require the same number of soldiers as 4 rows of 3 soldiers, but they are a different arrangement. Similarly if 4 people pay 3 pence each, the total amount paid is the same as that when 3 people pay 4 pence each.

20 POWERS OF NUMBERS

The product of equal numbers is called a **power**. Thus:

8×8 is called the second power of 8, or the square of 8.

$8 \times 8 \times 8$ is called the third power of 8, or the cube of 8, there being three equal factors.

$8 \times 8 \times 8 \times 8$ is called the fourth power of 8, there being four equal factors.

Similarly:

$a \times a$ is the second power of a or the square of a.
$a \times a \times a$ is the third   power of a or the cube of a.
$a \times a \times a \times a$ is the fourth power of a.

The process of writing a power in full is tedious, and the form of it restricts further operations, especially when the power is a high one. Accordingly mathematicians made many attempts through centuries to devise a symbolic method of representing the row of factors. Finally, Descartes in 1637 used a numeral to mark the number of factors or the power and wrote the cube of a for example as a^3, the fourth power as a^4, etc.

The figure used in this way is called an **index** or **exponent**; it indicates the number of factors.

Thus $a \times a \times a \times a \times a$ is written as a^5. With this symbolic method it is as easy to write down the 20th power of a number as the 2nd.

But a new symbol, if it is to be satisfactory, must not only express clearly and concisely the purpose for which it was devised, but it must also be convenient for operations with it. You will see, from what follows, that an index fulfils this condition, and later in the book you will see how it lends itself to important developments.

21 MULTIPLICATION OF POWERS OF A NUMBER

Suppose you require to multiply two powers of a – say $a^2 \times a^3$. These numbers written in full are

$$(a \times a) \times (a \times a \times a)$$

the brackets serving to separate the two powers.

It is a fundamental law of algebra, which will be assumed here, that when two groups of factors are multiplied, the factors in the groups are associated as one group of factors to give the product.

By this law
$$(a \times a) \times (a \times a \times a)$$
$$= (a \times a \times a \times a \times a).$$

∴ the number of factors in the product is the sum of the number of factors in the two groups.

∴ in the example above the product $(a \times a \times a \times a \times a)$ is the 5th power of a, and the index of the product is the sum of the indices of the two factors.

∴
$$a^2 \times a^3 = a^{2+3}$$
$$= a^5.$$

You can apply the same reasoning to other cases and deduce the general rule for the multiplication of two powers of a number.

When two powers of the same number are multiplied, the index of the product is the sum of the indices of the factors.

■ *Examples*

(1) $x^4 \times x^4 = x^{4+4} = x^8$.
(2) $2a^7 \times a^3 = 2a^{7+3} = 2a^{10}$.
(3) $5b^2 \times 3b^5 = 5 \times 3 \times b^2 \times b^5 = 15 \times b^{2+5} = 15b^7$.
(4) $a^2b \times ab^2 = a^2 \times a \times b \times b^2 = a^3b^3$.

This rule may be extended to the product of more than two factors. Thus:

$$x^2 \times x^3 \times x^4 = x^{2+3+4}$$
$$= x^9.$$

22 POWER OF A PRODUCT

To find the value of $(ab)^2$.

The use of the bracket shows that, as stated in Section 13, the expression within the bracket must be regarded as a whole.

∴ by definition of a power

$$(ab)^2 = (ab) \times (ab)$$
$$= (a \times a) \times (b \times b)$$
$$= a^2b^2.$$

You can see that the effect of this is that the index 2 must be distributed over each of the factors.

Thus $$(2xy)^3 = 2^3 \times x^3 \times y^3$$
$$= 8x^3y^3.$$

So in arithmetic $(2 \times 5)^2 = 2^2 \times 5^2 = 4 \times 25 = 100$. Consequently when taking a power of a product, the index of the power is said to be distributed and applied to each factor of the product.

EXERCISE 3

Write down the following in their simplest forms:

1 $4a \times 3$ **2** $5x \times 2y$

3 $\frac{1}{2}x \times 4y$ **4** $7m \times 3n$

5 $\frac{1}{3}a \times \frac{1}{4}b$ **6** $6a \times \frac{2}{3}b$

7 $3a \times 4b \times 5c$ **8** $\dfrac{x}{2} \times \dfrac{y}{3} \times \dfrac{z}{4}$

9 $x^2 \times x$ **10** $a \times a^2 \times a$

11 $x^2 \times x^2$ **12** $a^3 \times a^3$

13 $2a^2 \times a^3$

14 $3x^3 \times 2x^4$

15 $2ab \times ab$

16 $2b \times 3b^8$

17 $x^2y \times xy^2$

18 $7x^3a \times x^3a$

19 $2a \times 3a^2 \times a^3$

20 $(3a^2b)^3$

21 $(x^3)^2$

22 $(2a^4)^3$

23 $(2a^3)^4$

24 $(4a)^2 \times 4a^2$

Find the numerical values of the following:

25 $2a^2 \times a$, when $a = 3$

26 $2a^2 + a$, when $a = 3$

27 $a^2b \times ab^2$, when $a = 1$, $b = 2$

28 $x^2 + 7x + 2$, when $x = 10$

29 $3a^2 + 2ab + b^2$, when $a = 2$, $b = 3$

30 $c^2 \times c^3$, when $c = 1$

31 $3a \times 3a \times 3a$, when $a = 2$

32 $(2a^2x)^2$, when $a = 2$, $x = 3$

23 DIVISION OF POWERS

Suppose that a power of a number is divided by another power of the same number, as for example,

$$a^5 \div a^2.$$

Every division can be expressed in fractional form, as in arithmetic.

$\therefore$
$$a^5 \div a^2 = \frac{a^5}{a^2}$$

$$= \frac{a \times a \times a \times a \times a}{a \times a}$$

(by definition of a power).

As in arithmetic, you can cancel common factors in the numerator and denominator.

$\therefore$ you can cancel the two a factors in the denominator with two of the five factors of the numerator.

Then there will be left in the numerator $(5 - 2)$ factors.

$\therefore$
$$a^5 \div a^2 = a^{5-2}$$
$$= a^3.$$

Clearly you can follow the same method whatever the powers. Consequently you can deduce the rule:

When dividing a power of a number by another power of the same number, subtract the index of the divisor from the index of the dividend.

■ *Example:* Divide $84a^6$ by $12a^2$.

$$\frac{84a^6}{12a^2} = 7a^{6-2} = 7a^4.$$

■ *Example:* Divide $3x^4$ by $6x^6$.

Arranging as above,

$$3x^4 \div 6x^6 = \frac{3x^4}{6x^6}$$

$$= \frac{3 \times x \times x \times x \times x}{6 \times x \times x \times x \times x \times x \times x} .$$

In this example as the higher power is in the denominator, on cancelling there are $(6 - 4)$ factors in the denominator. Hence you get:

$$\frac{1}{2x^{6-4}} = \frac{1}{2x^2} .$$

EXERCISE 4

Write down answers to the following:

1 $a^3 \div a$	**2** $a^4 \div a^2$
3 $3x^3 \div x^2$	**4** $b^5 \div 2b$
5 $6a^6 \div 3a^2$	**6** $5y^4 \div y$
7 $6x^7 \div 2x^4$	**8** $14c^6 \div 7c^4$
9 $a^2b^3 \div ab^2$	**10** $x^5y^4 \div x^2y^3$
11 $5a^3b^3 \div ab^2$	**12** $6x^4y \div 2x^3$
13 $6a^4 \div 3a^4$	**14** $15x^3 \div 12x^3$

24 EASY FRACTIONS

Algebraic fractions obey the same fundamental laws as fractions in arithmetic. In principle they are manipulated by the same methods. But since the numerators and denominators may be algebraical expressions, sometimes rather complicated, they present difficulties not found in arithmetic fractions. In this chapter you will deal only with simple forms involving easy manipulation, more difficult cases being left until later.

25 ADDITION AND SUBTRACTION

You can apply the same methods as in arithmetic, as shown in the following examples.

■ *Example 1:* Find the sum of $\dfrac{x}{3} + \dfrac{x}{5}$.

This is of the same form as

$$\frac{2}{3} + \frac{2}{5}$$

and is worked in the same way.

Just as $\qquad \dfrac{2}{3} + \dfrac{2}{5} = \dfrac{(2 \times 5) + (2 \times 3)}{3 \times 5}$

so

$$\frac{x}{3} + \frac{x}{5} = \frac{(x \times 5) + (x \times 3)}{3 \times 5}$$

$$= \frac{5x + 3x}{15} = \frac{8x}{15} .$$

■ *Example 2:* Find the sum of $\dfrac{3}{a} + \dfrac{4}{b}$.

This is similar in type to the preceding example, and you deal with it in the same way:

$$\frac{3}{a} + \frac{4}{b} = \frac{(3 \times b) + (4 \times a)}{ab}$$

$$= \frac{3b + 4a}{ab}$$

■ *Example 3:* Simplify $\dfrac{x}{y} - \dfrac{a}{b}$.

Proceeding as before

$$\frac{x}{y} - \frac{a}{b} = \frac{(x \times b) - (a \times y)}{y \times b}$$

$$= \frac{bx - ay}{by} .$$

■ *Example 4:* Simplify $\dfrac{2a}{15} + \dfrac{5b}{12}$.

As in arithmetic, you find the L.C.M. of the denominators – namely 60.

Then
$$\frac{2a}{15} + \frac{5b}{12} = \frac{(2a \times 4) + (5b \times 5)}{60}$$

$$= \frac{8a + 25b}{60}.$$

Note: It is not possible to cancel any factors of 8 or 25 with factors of 60. This mistake is sometimes made by beginners, although the proceeding is contrary to the laws of arithmetic. Only factors common to each term of the numerator can be cancelled with factors of the denominator.

■ *Example 5:* Simplify $\frac{x}{12a^2b} - \frac{y}{18ab^2}$.

First find the L.C.M. of the denominators. To do this you find the L.C.M. of the numerical coefficients 12 and 18 – i.e., 36 – then the L.C.M. of a^2b and ab^2. This is a^2b^2, since both of them will divide into it exactly. The product of 36 and a^2b^2 is the L.C.M. of the denominator.

$$\frac{x}{12a^2b} - \frac{y}{18ab^2} = \frac{(x \times 3b) - (y \times 2a)}{36a^2b^2}$$

$$= \frac{3bx - 2ay}{36a^2b^2}.$$

26 MULTIPLICATION AND DIVISION

As in arithmetic, these operations are based upon the same important rule of fractions, namely:

If numerator and denominator are divided by the same number, the value of the fraction is unaltered.

You can express this algebraically as follows:

Let $\frac{a}{b}$ be any fraction.

Then
$$\frac{a}{b} = \frac{a \times n}{b \times n}$$

and
$$\frac{a}{b} = \frac{a \div m}{a \div m},$$

where m and n are any numbers.

■ *Example 1:* Simplify $\frac{4x^3y}{6xy^3}$.

Written in full the fraction is:

$$\frac{4x^3y}{6xy^3} = \frac{4 \times x \times x \times x \times y}{6 \times x \times y \times y \times y}.$$

Cancelling common factors, this is equal to:

$$\frac{2 \times x \times x}{3 \times y \times y} = \frac{2x^2}{3y^2}.$$

Note: In practice you do not need to write out the powers in full as shown above. You can apply the rule for the division of powers directly.

■ *Example 2:* Simplify $\dfrac{6ax^4}{14x^2y^2} \times \dfrac{2y^3}{3a^4}$.

As in arithmetic, you can cancel factors in either numerator with factors in either denominator.

$$\therefore \qquad \frac{6ax^4}{14x^2y^2} \times \frac{2y^3}{3a^4}.$$

$$= \frac{x^{4-2}}{7} \times \frac{2y^{3-2}}{a^{4-1}}.$$

$$= \frac{x^2 \times 2y}{7 \times a^3}$$

$$= \frac{2x^2y}{7a^3}.$$

■ *Example 3:* Simplify $\dfrac{8x^3}{5a^2y} \div \dfrac{4x^2}{3a}$.

Proceeding as with a similar arithmetical example:

$$\frac{8x^3}{5a^2y} \div \frac{4x^2}{3a} = \frac{8x^3}{5a^2y} \times \frac{3a}{4x^2}$$

$$= \frac{2x \times 3}{5ay}$$

$$= \frac{6x}{5ay}.$$

EXERCISE 5

Simplify the following:

1 $\dfrac{a}{5} + \dfrac{a}{7}$ **2** $\dfrac{2x}{3} - \dfrac{3x}{5}$

3 $\dfrac{4}{x} + \dfrac{5}{y}$

4 $\dfrac{3a}{2b} + \dfrac{5a}{3b}$

5 $\dfrac{9x}{10y} + \dfrac{5x}{8y}$

6 $\dfrac{7x}{3y} + \dfrac{2x}{9y} - \dfrac{5x}{6y}$

7 $\dfrac{5a}{6bc^2} + \dfrac{3b}{8a^2c}$

8 $x + \dfrac{4}{3x}$

9 $3ab + \dfrac{5}{2b}$

10 $8 - \dfrac{6}{5y^2}$

11 $a + 1 + \dfrac{1}{a}$

12 $x - 1 + \dfrac{1}{x}$

13 $\dfrac{1}{ab} + \dfrac{2}{bc} + \dfrac{3}{ac}$

14 $\dfrac{5}{a^2bc} - \dfrac{4}{ab^2c} + \dfrac{2}{abc^2}$

15 $\dfrac{x^4y^2}{x^2y}$

16 $\dfrac{12a^2b}{3a^4}$

17 $\dfrac{15a^2b^4c}{10a^4b^2c^2}$

18 $\dfrac{x^3 \times x^3}{x^6}$

19 $\dfrac{4x^4 \times 2x^3y}{8x^6}$

20 $\dfrac{a^4 \times bc^2}{a^3 \times b^2c^3}$

21 $\dfrac{2a}{9} \times \dfrac{6}{a^2}$

22 $\dfrac{x^2}{xy} \times \dfrac{3y^2}{2x}$

23 $\dfrac{3p}{2q^2} \times \dfrac{6q}{5p}$

24 $2x^2 \times \dfrac{y}{3x}$

25 $2x \times \dfrac{x}{y}$

26 $2x \div \dfrac{x}{y}$

27 $\dfrac{x}{y} \div \dfrac{x}{y}$

28 $\dfrac{x}{y} \div \dfrac{y}{x}$

29 $\dfrac{3a}{7} \div \dfrac{9a}{14}$

30 $\dfrac{2x^2}{y} \div \dfrac{3y}{x^2}$

31 $\dfrac{6x^2}{8yz} \div \dfrac{3xy}{4xz}$

32 $\dfrac{a^5}{b^4} \div \dfrac{3a^2}{2b^2}$

33 $\dfrac{4abc}{5b^2} \div \dfrac{3c^2}{10ab}$

34 $3xyz \div \dfrac{4xz}{y}$

35 $30 \div \dfrac{3xyz}{2z}$

36 $\dfrac{ab}{c^2} \div \dfrac{c^2}{4ab}$

3

Brackets and Operations with Them

27 REMOVAL OF BRACKETS

Simple examples of the use of brackets as a convenient way of grouping numbers have already been considered. In this chapter extensions of their use, operations with them and the simplification of algebraical expressions, which contain brackets, by removing them are all examined.

We will begin with a simple but important case.

Fig. 5 represents a rectangle made up of two other rectangles.

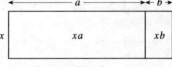

Fig. 5

Let a mm = the length of one rectangle

Let b mm = the length of the other rectangle

Let x mm = the breadth of each rectangle as shown in Fig. 5.

Then $(a + b)$ mm = the length of the combined rectangle, placing $(a + b)$ in brackets shown in Section 13.

The areas of two smaller rectangles are xa and xb mm^2 and area of whole rectangle = $x(a + b)$ mm^2.

But the area of the whole rectangle equals the sum of the area of the parts.

$$\therefore \qquad x(a + b) = xa + xb.$$

Similarly if there are three rectangles of lengths a, b and c mm respectively, then in the same way you could show that:

$$x(a + b + c) = xa + xb + xc.$$

By modifying the figures, with a little ingenuity, you could similarly show that:

(1) $$x(a - b) = xa - xb$$
(2) $$x(a + b - c) = xa + xb - xc$$
(3) $$x(a - b + c) = xa - xb + xc.$$

In all these examples expressions containing brackets have been transformed into expressions without brackets, or, alternatively, the brackets have been removed.

Hence you can deduce that:

When the whole of an expression within brackets is multiplied by any number, then, if the brackets are removed, each term within the brackets must be multiplied by the number.

The factor without the brackets is said to be distributed as a factor of each term within the brackets.

This is an example of the algebraic law called the Law of distribution.

28 ADDITION AND SUBTRACTION OF EXPRESSIONS WITHIN BRACKETS

Here are four cases to consider. They are represented by the following expressions:

(1) $a + (b + c)$.
(2) $a + (b - c)$.
(3) $a - (b + c)$.
(4) $a - (b - c)$.

The question to be considered is, What will be the effect of removing the brackets in the above expressions, a, b and c representing any numbers?

(1) $a + (b + c)$

Let a, b, c be represented by the areas of rectangles as shown in Fig. 6.

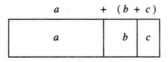

Fig. 6

It is evident that the area of the whole rectangle which represents $a + (b + c)$ is the sum of the three rectangles representing a, b, c.

$$\therefore \qquad a + (b + c) = a + b + c$$

The steps in the addition of the numbers in this case are not altered by inserting or removing the brackets.

(2) $a + (b - c)$

As in the previous case, a, b, c are represented by the areas of rectangles, now as in Fig. 7.

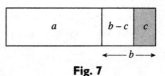

Fig. 7

The two unshaded rectangles represent a and $b - c$.

$\therefore$ the whole unshaded portion represents $a + (b - c)$. You can form it either:

(1) by adding $(b - c)$ to a.
(2) by adding b to a and then subtracting c.

These two results are equal.

$$\therefore \qquad a + (b - c) = a + b - c$$

Thus no change results when the brackets are removed.

(3) $a - (b + c)$

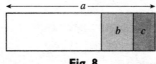

Fig. 8

Using the same method as before of representing a, b and c, the unshaded rectangle (Fig. 8) represents the remainder when $(b + c)$, the two shaded rectangles, are subtracted from a, the whole rectangle, i.e., it represents $a - (b + c)$.

Also if from a, the whole rectangle, you subtract b and c in turn, the remaining rectangle is the unshaded portion.

$\therefore$ it represents $a - b - c$.

$$\therefore \qquad a - (b + c) = a - b - c.$$

(4) $a - (b - c)$

In Fig. 9 the rectangle representing a, b, c is shown, a being represented by the whole rectangle.

The shaded rectangle represents $(b - c)$.

The unshaded rectangle represents the result of subtracting $(b - c)$ from a, i.e., it represents $a - (b - c)$.

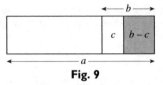

Fig. 9

It may also be considered as representing the result of subtracting b from a and then adding c, i.e. it represents $a - b + c$.

$$\therefore \qquad a - (b - c) = a - b + c.$$

Collecting the four cases you have:

(1) $a + (b + c) = a + b + c$.
(2) $a + (b - c) = a + b - c$.
(3) $a - (b + c) = a - b - c$.
(4) $a - (b - c) = a - b + c$.

From these results you can deduce two rules respecting signs when the brackets are removed.

A. *From (1) and (2) when the $+$ sign precedes the brackets the signs of the terms within the brackets are unaltered.*

B. *From (3) and (4) when the $-$ sign precedes the brackets the signs of the terms within the brackets are changed.*

29 WORKED EXAMPLES

The following are examples of the use made of the rules of Section 27 and Section 28 when brackets are removed from an algebraic expression in order to simplify it.

■ *Example 1:* Simplify $a(a^2 + ab + b^2)$.

When you remove the brackets you use the rule of Section 27, you multiply each term within the brackets by the factor a.

Thus: $\qquad a(a^2 + ab + b^2) = a^3 + a^2b + ab^2$.

■ *Example 2:* Simplify $2(4a + 3b) + 6(2a - b)$.

When removing the brackets you use Section 27 to multiply by 2 and 6, and from Section 28, since the $+$ sign before the second pair of brackets is positive, there is no change of sign.

$$2(4a + 3b) + 6(2a - b)$$
$$= 8a + 6b + 12a - 6b$$
$$= 20a.$$

■ *Example 3:* Simplify $5x - (5y + 2x)$.

This is an example of case (3) of Section 28. On removing the brackets signs are changed.

$$
\begin{aligned}
\text{Thus} \quad & 5x - (5y + 2x) \\
= & 5x - 5y - 2x \\
= & 3x - 5y \text{ (on adding like terms).}
\end{aligned}
$$

■ *Example 4:* Simplify $3(4a - b) - 2(3a - 2b)$.

This involves the rule of Section 27 and case (4) of Section 28. Using these

$$
\begin{aligned}
& 3(4a - b) - 2(3a - 2b) \\
= & 12a - 3b - 6a + 4b \\
= & 6a + b.
\end{aligned}
$$

■ *Example 5:* Simplify $x(2x - y) - x(x - y) - y(x + 2y)$ and find its value when $x = 2$, $y = 1$.

$$
\begin{aligned}
& x(2x - y) - x(x - y) - y(x + 2y) \\
= & 2x^2 - xy - x^2 + xy - xy - 2y^2 \\
= & x^2 - xy - 2y^2 \text{ (since } +xy - xy = 0).
\end{aligned}
$$

Substituting $x = 2$, $y = 1$.

$$
\begin{aligned}
x^2 - xy - 2y^2 = & (2)^2 - (2 \times 1) - 2(1)^2 \\
= & 4 - 2 - 2 \\
= & 0.
\end{aligned}
$$

Note: This example shows the advantage of simplifying the expression before substituting the values of x and y.

EXERCISE 6

Simplify the following expressions by removing brackets.

1 $3(5x + 6z)$
2 $2a(3a + 4b)$
3 $6a^2(3a + 7b - 6c)$
4 $2(x + 2y) + 3(2x - y)$
5 $x(x^2 - 3x) + x^2(4x + 7)$
6 $\frac{1}{2}(x - 2y) + \frac{1}{4}(2x + 4y)$
7 $2(x + y + z) + 3(2x + y - 2z)$
8 $x - (2y + z)$
9 $2x - (y - 2z)$
10 $2(2a + 2b) - 3(a - b)$

11 $3a - (2a + b)$
12 $3a - (2a - b)$
13 $5x - (x - 2y + 2z)$
14 $3(a + b - c) - 2(a - b + c)$
15 $4(x + y) - 3(2x - y) + 2(x - 2y)$
16 $a(a + b) - b(a - b)$
17 $x^2(x + y) - xy(x^2 - y^2)$
18 $3(x^2 + x + 5) - 2(x^2 - 3x - 4)$
19 $2p(3p + 2q) - 3q(2p - 5q) + p(3p + 5q)$
20 $5(xy)^2 - 3x(y - 2x)$
21 $(2x^2)^2 - 2x^3(x - 4)$

30 SYSTEMS OF BRACKETS

It may happen that an expression within brackets is part of another expression which is itself within brackets. In that case a second set of brackets would be required, and to avoid confusion they must be a shape different from those already used, such as {—} or [—].
   For example:

$$40 - \{2(a + b) + 5(a - b)\}.$$

You will easily recognise how clearly and effectively the brackets help to show the construction of the expression and relations of the different parts to one another.
   It might happen that the whole of the above expression is to be multiplied by $2b$. This will necessitate another set of brackets which will indicate that the expression is to be treated as a whole. You would express this as follows:

$$2b[40 - \{2(a + b) + 5(a - b)\}].$$

Thus $\qquad\qquad 5a - \dfrac{a - b}{2}$

means the same as $\qquad 5a - \frac{1}{2}(a - b).$

   If the whole expression were multiplied by 2, it would become $10a - (a - b)$.
   When expressions with two or more sets of brackets are to be simplified by removal of the brackets, it is best, as a rule, to begin with the inside brackets and work outwards. You will see this in the following examples.

31 WORKED EXAMPLES

■ *Example 1:* Simplify $2\{3a + 5(b + c)\}$.

As stated above you begin by removing the inner brackets.

$$2\{3a + 5(b + c)\}$$
$$= 2\{3a + 5b + 5c\}$$
$$= 6a + 10b + 10c.$$

■ *Example 2:* Simplify $3\{3a - 2(a - b)\}$.

$$3\{3a - 2(a - b)\}$$
$$= 3\{3a - 2a + 2b\}$$
$$= 3\{a + 2b\}$$
$$= 3a + 6b.$$

■ *Example 3:* Simplify $12a - 2[3a - \{4 - 2(a - 3)\}]$.

Beginning with the innermost bracket

$$12a - 2[3a - \{4 - 2(a - 3)\}]$$
$$= 12a - 2[3a - \{4 - 2a + 6\}]$$
$$= 12a - 2[3a - \{10 - 2a\}] \text{ (adding like terms)}$$
$$= 12a - 2[3a - 10 + 2a]$$
$$= 12a - 2[5a - 10]$$
$$= 12a - 10a + 20$$
$$= 2a + 20.$$

EXERCISE 7

Remove the brackets from the following expressions and simplify them.

1 $3\{5a - 3(a + 1)\}$
2 $3\{4(a + b) - 3(a - 2b)\}$
3 $\frac{1}{2}\{6x - 3(2x - 1)\}$
4 $5a^2 + 2a\{b - (a + c)\}$
5 $3p^2 - \{2p^2 - p(p + 1)\}$
6 $3x(x + 3y) - 2\{x^2 + 3y(x - 2y)\}$
7 $3bc - 2\{b(b - c) - c(b + c)\}$
8 $15x - [3x - \{2x - (x - 5)\}]$
9 $50 - 2[3a + 2\{3b - 4(b - 1)\}]$
10 $2(x + y) - \overline{x - y}$

11 $12\left\{\dfrac{a-b}{2}-\dfrac{a+b}{3}\right\}$

12 $3x\left\{\dfrac{2x-y}{3}-\dfrac{x-2y}{6}\right\}$

13 $3c-\left\{\dfrac{2a+c}{8}-\dfrac{a+2c}{4}\right\}$

Fill in the blanks within the brackets in the following:

14 $2a-b+c=2a-($ $)$
15 $x-y-z=x-($ $)$
16 $2a+4b-6c=2($ $)$
17 $x^2-xy+y^2=x^2-y($ $)$

18 From $3a-2b+4c$ subtract $a+2b-3c$
19 Take $2x-3y+4z$ from $3x-y+2z$
20 When $a=3$, $b=2$, $c=1$, find the values of the following:
    (1) $4a(a+4b)-a(3a-b)$
    (2) $3c\{4c-(3c-1)\}$

4

Positive and Negative Numbers

32 THE SCALE OF A THERMOMETER

Figs. 10a and 10b represent portions of Celsius thermometers in which a fine column of mercury registers the rise and fall of temperature.

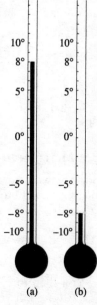

The zero point, marked 0, indicates the position of the mercury in the tube at freezing point – i.e., the freezing point of water.

Fig. 10(a) shows the mercury at 8°C above zero.

Now suppose the temperature falls 16°C below this point.

First it falls 8°C to 0°C, and then continues to fall for 8°C below zero. To show this temperature on the scale it must be marked in some way which is different from the 8°C above zero, or there would be confusion. To distinguish the degrees below zero from those above a minus sign, −, is put before all those below zero, and if necessary a plus sign, +, before those above zero.

Thus +8°C means 8 degrees above zero, and −8°C means 8 degrees below zero.

These are called positive and negative degrees, and the signs + and − are used to indicate different directions up and down from the zero.

(a)   (b)

Fig. 10

33 An example from time

In reckoning time, the numbers denoting the years may be counted from a particluar date, e.g. the birth of Christ. Years after that event are denoted by AD, as AD 1941, and those before by BC, as 55 BC. The use of the symbols AD and BC is, in principle, similar to the use of + and − in the case of the thermometer.

34 A commercial illustration

In a certain transaction at market a farmer made a profit of £12.

In a second deal the farmer lost £8.

Consequently in the two transactions the farmer made a net gain of £12 − £8 = £4.

In a third transaction he lost £10.

The farmer's profit and loss account is now shown by £4 − £10.

If this loss had been £4, his position would be £4 − 4 = 0, i.e., a zero position, neither loss nor gain.

But the farmer's loss is £10, not £4; therefore the net loss is of £6, i.e., the farmer has £6 below zero.

To distinguish gains from losses, you could, as in the case of the thermometer, place the

> negative sign before amounts showing losses,

and the

> positive sign before amounts showing profits.

In that sense £4 − £10 = −£6, the negative sign indicating a loss of £6.

35 Motion in opposite directions

Suppose you start from a point O (see Fig. 11) travel for 4 km in the direction O to X, reaching the point marked A.

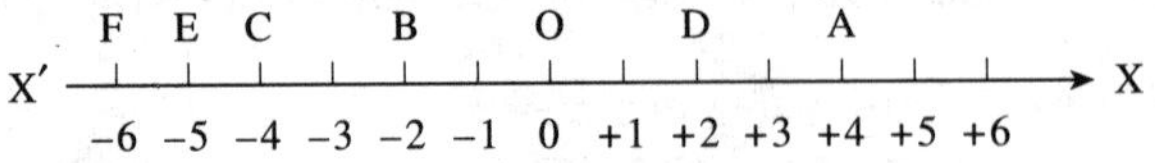

Fig. 11

You then turn and travel 6 km in the opposite direction, O to X′. After 4 km you reach O, the zero or starting point. The next 2 km take you to B.

You are now 2 km from O but in the direction opposite to that in which you started. Your successive distances from O can be shown by +4 − 6. This suggests that, as in the previous cases, if distances from O in one direction were regarded as positive, the distances in the opposite direction could be regarded as negative.

Thus, if you now say that you are −2 km from O, you mean that you are 2 km in the direction opposite to the original. Accordingly, in the diagram showing the movements from O (Fig. 11).

Distances to the right with + signs are called positive.
Distances to the left with − signs are called negative.

With this device, when giving your position from O, the sign of the number would indicate in which direction you are from O.

Thus −4 km would indicate you are at C, +2 km would show you are at D.

The number with the + sign is called a positive number.
The number with the − sign is called a negative number.

36 POSITIVE AND NEGATIVE NUMBERS

From this it appears that a new kind of number has been devised – namely, a negative number – and that, in consequence, you can divide numbers into two kinds: positive and negative. From the examples above a negative number is a number which in its meaning and effect is opposite to a positive number.

Frequently, as in examples of Sections 32 and 35, the negative number indicates a direction opposite to that of the positive number, and in this sense.

positive and negative numbers are called **directed** numbers.

If negative numbers can rightly be classed as numbers, they must, in operations with them, conform to the rules governing the numbers which we now call positive numbers. These operations will be considered fully later, but a few simple illustrations will serve to show that you can deal with them in the same way as positive numbers.

For example, in the matter of addition, you can add −2 and −3, and a glance at Fig. 11 will show that the result is −5, being equal to the sum of −2 and another −3 from O to the point marked E.

Or if −3 is multiplied by 2, i.e., you double the distance from O to get −6, at the point F.

Similarly division of −6 by 2 would give −3.

For the rest of this chapter, in order to make the meaning clear, positive and negative numbers, when being used in operations will be placed in brackets.

Thus $(-6) \div (+2) = (-3)$.

37 NEGATIVE NUMBERS

Corresponding to every positive number there is a negative number, and you can write a sequence of negative numbers corresponding to positive numbers.

Thus if you write down the numbers beginning, for example, with +6, and decreasing by one at each step, you get the sequence of numbers +6, +5, +4, +3, +2, +1, 0.

With the negative number, you do not stop at the zero, but continue with the subtraction, so that you get

$$-1, -2, -3, -4, -5 \ldots, \text{etc.}$$

in descending order of magnitude. Or if you start with (-6) and add unity in succession you get the complete sequence:

$$-6, -5, -4, -3, -2, -1, 0, +1, +2, +3, +4, +5 \ldots,$$

in ascending order of magnitude. You can extend this sequence in either direction and decimals and fractions fall into their places between these numbers. Thus you get what is called the **number line**.

38 GRAPHIC REPRESENTATION OF THE NUMBER LINE

The graphical representation of the number line is so important that we return to it again.

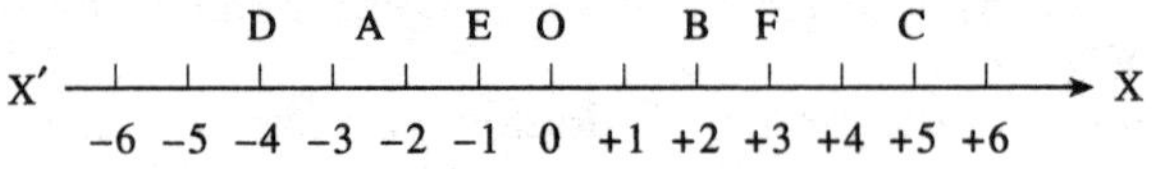

Fig. 12

The straight line XOX′ is drawn, as in Fig. 12, to represent a small part of the line.

On this line, starting from a point O, and using a suitable scale, distances are marked to the right to represent positive numbers and to the left to represent negative numbers.

You could imagine this line to be extended to any distance on either side so that any number could be included. Numbers involving decimals lie between those marked. Thus -2.5 would be at A.

Two principles may be noted:

(1) every number can be represented at its appropriate point on the scale;
(2) conversely, every point on the scale represents a number.

You should observe that the numbers represented in the figure increase from left to right, as shown by the arrow.

39 Operations with negative numbers

With the introduction of negative numbers, algebra passes beyond the boundaries of arithmetic. It is necessary therefore to examine the operations of addition, subtraction, multiplication and division with this new kind of number, remembering, as previously stated, that it must conform to the laws of algebra.

40 Addition of positive and negative numbers

You have already seen in Section 36 that addition of two negative numbers is performed in the same way as that of positive numbers.

Just as $\qquad$ $(+3) + (+2) = +5$
so $\qquad$ $(-3) + (-2) = -5$.

Such operations can be confirmed by use of Fig. 12.

The addition of a positive and a negative number can also be seen from Fig. 12.

For example, $(-4) + (+3)$ is represented by starting at D, which represents -4, and moving $+3$ to the right to E, the result being -1.

Similarly $(+3) + (-7)$ as found by starting at F, marking $+3$, and since -7 is a negative number, you move 7 divisions to the left to D to find the sum, which is (-4).

When the negative numbers involve letters, the procedure is the same. Thus:

$$(-5a) + (+a) = -4a$$
$$(+2b) + (-5b) = -3b$$
$$(-2x) + (+6x) = 4x.$$

41 SUBTRACTION

This operation presents a little more difficulty, since it is not easy at first to understand what is implied by the subtraction of a negative number, as, for example, $(+6) - (-2)$ or $(-2) - (-5)$.

This can be deduced from Fig. 12, but the rule will first be obtained by applying a fundamental law of addition and subtraction.

Since $9 = 7 + 2$
then $9 - 2 = 7$
and $9 - 7 = 2,$

or, in general terms: if

$$a = b + c$$
then $a - b = c$
and $a - c = b.$

You have seen above that

$$(-5) + (+3) = -2$$
or $(-2) = (-5) + (+3).$

∴ from the above

$$(-2) - (-5) = +3$$
but you know that $(-2) + (+5) = +3.$

∴ comparing the two statements

$$-(-5) = (+5).$$

A similar result will clearly hold whatever the numbers.
∴ You can conclude that for any number a

$$-(-a) = +a.$$
Similarly $-(-2a) = +2a.$

■ *Examples:*

$$5x - (-3x) = 5x + 3x = 8x$$
$$-2b - (-4b) = -2b + 4b = 2b.$$

Graphical illustrations

To find $(-2) - (-5)$,

You can deduce the rule from the graphical representation of the number scale in Fig. 12 as follows:

If you add a negative number you move to the left along the scale.

Thus $$(-2) + (-5) = -7.$$

Consequently if you subtract a negative number you must move to the right.

Starting from (-2) and moving 5 to the right you reach $+3$, i.e.,

$$(-2) - (-5) = (+3).$$

To find $(-2) - (+5)$.

When adding a positive number you move to the right.

∴ when subtracting a positive number you move to the left.

∴ starting from (-2) you move 5 divisions to the left and read (-7).

Summarizing the rules for addition and subtraction:

$$+(+a) = (+a)$$
$$+(-a) = (-a)$$
$$-(+a) = (-a)$$
$$-(-a) = (+a)$$

You should compare these with rules for signs given in Section 28.

EXERCISE 8

1   A lift starting from the ground-floor rises to the fourth floor. Then it descends to the second floor, rises to the sixth floor and finally descends to the ground-floor. Express its movements by using positive and negative numbers.

2   The movement of the mercury in a thermometer was as follows. Starting at $+8°C$ it rose $2°C$, fell $14°C$, then rose $4°C$ and finally fell $6°C$. Express these, using positive and negative signs, and find the final temperature.

3   How much higher than a temperature of $-15°C$ is:

(1) A temperature of $-4°C$?
(2) Freezing point?
(3) $+15°C$?

4 Using the number scale shown in Fig. 12 find:

  (1) By how much -2 is greater than -7?
  (2) By how much -6 is less than -1?
  (3) By how much $+3$ is greater than -5?

5 (1) What must be added to (-3) to give (a) -1, (b) $+1$?
  (2) What must be taken from (-3) to give (-8)?

6 Write down the values of:

  (1) $(+6) + (-2)$ (2) $(+6) - (-2)$
  (3) $(-6) + (-2)$ (4) $(-6) - (-2)$
  (5) $0 - (-3)$ (6) $(-4) + (-4)$
  (7) $-(4) - (-4)$ (8) $(-4) - (+4)$

7 Simplify the following:

  (1) $+2a - (-5a)$
  (2) $-4x - (+3x)$
  (3) $+3ab - (-7ab)$
  (4) $2x - 3y - 5y - 3x$
  (5) $(3a - 2b) - (2a + 5b)$
  (6) $(3x - y) - (4x - 3y)$
  (7) $3x - (3y - 4x)$
  (8) $(5 + x) - (6 - 2x) - (3x + 7)$

8 (1) Subtract $(x - 2y)$ from $(3x - 4y)$.
  (2) Subtract $(x - y + 2z)$ from $(3x - 2y - 5z)$.

9 Fill in the brackets in the following:

  (1) $3a - ($ $) = 8a$
  (2) $5x - ($ $) = -x$
  (3) $-3a - ($ $) = 7a$

10 Write down the values of:

  (1) $0 - (+a)$ (2) $0 - (-a)$
  (3) $0 + (-a)$

---

42 MULTIPLICATION

(1) Multiplication of $(-a)$ by $(+b)$
and $(+a)$ by $(-b)$.

Consider, as a special case $(-2) \times (+3)$.

Since multiplication is a shortened form of addition, the meaning of $(-2) \times (+3)$ is $(-2) + (-2) + (-2)$.
This $= (-6)$.

$$\therefore \qquad (-2) \times (+3) = (-6).$$

You can apply this to any pair of numbers, so you can conclude that in general

$$(-a) \times (+b) = (-ab).$$

Since the multiplication of two numbers can be taken in any order (Section 19)

$$(+a) \times (-b) = (-b) \times (+a)$$

and this, by the above result, is $(-ab)$.

(2) Multiplication of $(-a)$ by $(-b)$.

Since $\qquad (-a) \times (+b) = (-ab)$

and a negative number operates in the opposite sense to a positive number, it follows that

$$(-a) \times (-b) = (+ab).$$

43 DIVISION

(1) Division of $(+a)$ by $(+b)$.

Since $\qquad (+4) \times (+3) = +12,$
$\therefore \qquad (+12) \div (+4) = (+3).$

Similarly $\qquad +(a) \div (+b) = +\left(\dfrac{a}{b}\right).$

(2) Division of $(-a)$ by $(+b)$.

Since $\qquad (+4) \times (-3) = (-12),$
$\therefore \qquad (-12) \div (+4) = (-3).$

Similarly $\qquad (-a) \div (+b) = \left(-\dfrac{a}{b}\right).$

(3) Division of $(+a)$ by $(-b)$.

Again $\qquad (-4) \times (-3) = (+12),$
and $\qquad (+12) \div (-4) = (-3).$

Similarly $\qquad (+a) \div (-b) = \left(-\dfrac{a}{b}\right).$

(4) Division of $(-a)$ by $(-b)$.

As above $\qquad\qquad (+4) \times (-3) = (-12),$
$\therefore \qquad\qquad\quad (-12) \div (-3) = (+4),$

and in general $\qquad (-a) \div (-b) = \left(+\dfrac{a}{b}\right).$

44 SUMMARY OF RULES OF SIGNS FOR MULTIPLICATION AND DIVISION

Multiplication

$$(+a) \times (+b) = +ab$$
$$(+a) \times (-b) = -ab$$
$$(-a) \times (+b) = -ab$$
$$(-a) \times (-b) = +ab$$

Division

$$(+a) \div (+b) = \left(+\frac{a}{b}\right)$$

$$(+a) \div (-b) = \left(-\frac{a}{b}\right)$$

$$(-a) \div (+b) = \left(-\frac{a}{b}\right)$$

$$(-a) \div (-b) = \left(+\frac{a}{b}\right)$$

These results can be summarised in the following rule:

In the multiplication and division of positive and negative numbers, if the two numbers have the same sign the result is a positive number. If the signs are different, the result is a negative number.

Or to remember these more readily the following slogan can be used:

Like signs give $+$
Unlike signs give $-$

45 POWERS, SQUARES AND SQUARE ROOTS

When you square a number you multiply two numbers with the same signs. In accordance with the above rules, the product must be positive.

Thus
$$(+a) \times (+a) = +a^2$$
$$(-a) \times (-a) = +a^2.$$

Consequently the square of any number is positive or zero.

It follows that when this operation is reversed and the square root of a^2 is required, this may be either $(+a)$ or $(-a)$.

To indicate this we use the sign $\pm$, meaning 'plus or minus', i.e.,

$$\sqrt{a^2} = \pm a.$$

Again

$$(-a)^3 = (-a) \times (-a) \times (-a) = -a^3$$

and

$$(-a)^4 = (-a) \times (-a) \times (-a) \times (-a) = +a^4$$

From these and similar examples you can deduce that:

An odd power of a negative number is negative.
An even power of a negative number is positive.

EXERCISE 9

1 Write down the answers to the following:

(1) $(+12) \times (+3)$ (2) $(+12) \times (-3)$
(3) $(-12) \times (+3)$ (4) $(-12) \times (-3)$
(5) $(+12) \div (+3)$ (6) $(+12) \div (-3)$
(7) $(-12) \div (+3)$ (8) $(-12) \div (-3)$

2 Write down the answers to the following:

(1) $(+a) \times (-a)$ (2) $(-a) \times (-a)$
(3) $(+a) \div (-a)$ (4) $(-a) \div (-a)$

3 Write down the answers to the following:

(1) $(-2a) \times (+2b)$ (2) $(-2a) \times (-2b)$
(3) $(+10x) \times (-2y)$ (4) $(-10x) \times (+2y)$
(5) $(+10x) \div (-2y)$ (6) $(-10x) \div (-2y)$

4 Find the values of the following:

(1) $(-4) \times (+3) \times (-2)$
(2) $(-a) \times (+3a) \times (-2a)$
(3) $(-18xy) \div (-6x)$
(4) $(-24a^2b^2) \div (+4ab)$

5   Find the values of the following:

    (1) $(-5x) \times (-2x) \times (-x)$
    (2) $a(a - b) - b(b - a)$
    (3) $-\{a(-2b) \times (-b)\}$
    (4) $-a(a - 2b - c)$

6   Find the simplified form of the following:

    (1) $(+2a) \times (-5b) \times (-2b)$
    (2) $(-4x)^2 - 2x(5x - 4)$
    (3) $x(y - z) - z(x - y) - y(x - z)$

7   Write down the second, third, fourth and fifth powers of:

    (1) $(-a)$ (2) $(-2x)$ (3) $\left(-\dfrac{b}{3}\right)$

8   Write down the square roots of 81 and $9x^4$ and the cube roots of $-x^3$ and $-8a^6$.

9   Find the answers to the following:

    (1) $(-8x) \times (-2)$
    (2) $(-10x) \div (-2)$
    (3) $(-2xy) \div (-x)$
    (4) $(+6b) \div (-3)$
    (5) $(+8t^2) \div (-4t)$
    (6) $(-4x^3) \div (-2x^2)$
    (7) $(-2x^2) \times (-4x)$
    (8) $(+15x^2y) \div (-5xy)$
    (9) $(-12a^2b^2) \div (+3ab)$
    (10) $(-24a^3bc^2) \div (-4abc)$

10   Write down the values of:

    (1) $\{(-a) \times (-b)\} \div (-a)$

    (2) $\left(+\dfrac{1}{x^2}\right) \times (-x)^3$

    (3) $(-6x)^2 \times (-x)^3 \div (-2x)^4$

5

Expressions and Equations

46 UNDERSTANDING EXPRESSIONS

Many situations can be described using letters as a shorthand. For example when you find the area of a rectangle you multiply the length by the breadth. If you use l to represent the length and b to represent the breadth then the area can be written as $l \times b$ or more simply by using the expression lb.

In algebra you often use the letter n to represent the position of the nth term in a sequence of numbers. An expression is then given which shows how you can find the value of the term for a particular value of n.

■ *Examples*: The nth term in the sequence 2, 4, 6, 8, 10, ... is $2n$. Here the 5th term is 2×5 i.e. 10.
The nth term in the sequence 1, 4, 9, 16, 25, ... is n^2.
Here the 5th term is 5×5 i.e. 25.

Often the expressions will be much more complicated than these and you will need to understand exactly what the expression means before you can use it to find the value of a particular term in the sequence.

You saw on page 9 in Chapter 1 how you could use the expression $3n + 1$ to describe the nth term in the sequence of numbers 4, 7, 10, 13. ... You find the 5th number in this sequence by substituting 5 as the value of n in the expression, i.e. $3 \times 5 + 1$. The 5th number is therefore 16.

The expression $3n + 1$ means:

first multiply the number n by 3 and then add 1.

A helpful way to illustrate this expression is to draw a diagram.

number → ⬚multiply by 3⬚ → ⬚add 1⬚ → 3 × number + 1

You can shorten this to:

n → ⬚× 3⬚ → ⬚+ 1⬚ → 3 × n + 1 or 3n + 1

A diagram like this is often called a function machine. The arrows are used to show the order in which the operations are carried out.

If you use a particular value of n, say 5, this is fed through the machine and if necessary you show the intermediate steps in the calculation.

5 → ⬚× 3⬚ → 15 → ⬚+ 1⬚ → 3 × 5 + 1 or 16

47 USING FUNCTION MACHINES

A function machine is not only a useful aid in helping you to write the correct expression to describe the nth term in a sequence, but it also helps you to carry out the arithmetic operations in the correct order.

■ *Example 1:* Draw a function machine to describe:

(a)   5, 8, 11, 14, 17 . . .
        i.e. *first multiply by* 3 *and then add* 2.
(b)   9, 12, 15, 18, 21 . . .
        i.e. *first add* 2 *and then multiply by* 3.

What is the result using 5 with each machine?

(a) n → ⬚× 3⬚ →   ⬚+ 2⬚ → 3 × n + 2 or 3n + 2
       5 → ⬚× 3⬚ → 15 →   ⬚+ 2⬚ → 17

(b) n → ⬚+ 2⬚ →   ⬚× 3⬚ → (n + 2) × 3 or 3(n + 2)
       5 → ⬚+ 2⬚ → 7 →   ⬚× 3⬚ → 21

You should notice that the order in which the operations are carried out affects both the actual form of the expression and also the numerical results. The brackets are needed in the second expression to show that the whole of n + 2 is multiplied by the 3.

■ *Example 2:* Draw a function machine to describe:

(a)　5, 12, 19, 26, 33 . . .
　　　i.e. *first multiply by* 7 *and then subtract* 2.
(b)　3, 6, 11, 18, 27 . . .
　　　i.e. *first square the number and then add* 2.

What is the result using 5 with each machine?

(a)　$n \rightarrow \boxed{\times 7} \rightarrow$　　　$\boxed{-2} \rightarrow 7 \times n - 2$ or $7n - 2$
　　　$5 \rightarrow \boxed{\times 7} \rightarrow 35 \rightarrow$　　$\boxed{-2} \rightarrow 33$

(b)　$n \rightarrow \boxed{\text{square}} \rightarrow$　　$\boxed{+2} \rightarrow n \times n + 2$ or $n^2 + 2$
　　　$5 \rightarrow \boxed{\text{square}} \rightarrow 25 \rightarrow$　$\boxed{+2} \rightarrow 27$

EXERCISE 10

1　Draw a function machine to describe:

　　　(*a*)　*first multiply by* 2 *and then add* 3
　　　(*b*)　*first multiply by* 6 *and then subtract* 5
　　　(*c*)　*first add* 3 *and then multiply by* 2
　　　(*d*)　*first square the number and then subtract* 2

　　In each of the above find the result of using 7 with the machine.

2　Describe how to find the *n*th term of the sequence and then draw the function machine for the corresponding expression.

　　　(*a*)　3, 6, 9, 12, 15 . . .　　　(*d*)　1, 4, 9, 16, 25 . . .
　　　(*b*)　2, 5, 8, 11, 14 . . .　　　(*e*)　11, 14, 19, 26, 35 . . .
　　　(*c*)　1, 4, 7, 10, 13 . . .　　　(*f*)　6, 11, 16, 21, 26 . . .

■ *Example 3:* Draw a function machine to show how to evaluate the expression: (a) $5n - 2$　(b) $4(n + 5)$　(c) $n^2 + 5$

In each case write down the first five terms of the corresponding sequence:

(a) $n \rightarrow \boxed{\times 5} \rightarrow \boxed{-2} \rightarrow 5 \times n - 2$ or $5n - 2$

　$1 \rightarrow 3, 2 \rightarrow 8, 3 \rightarrow 13, 4 \rightarrow 18, 5 \rightarrow 23$
　so the first five terms of the sequence are 3, 8, 13, 18 and 23.

(b) $n \rightarrow \boxed{+5} \rightarrow \boxed{\times 4} \rightarrow (n + 5) \times 4$ or $4(n + 5)$

　$1 \rightarrow 24, 2 \rightarrow 28, 3 \rightarrow 32, 4 \rightarrow 36, 5 \rightarrow 40$
　so the first five terms of the sequence are 24, 28, 32, 36 and 40.

(c) $n \rightarrow$ ☐square☐ $\rightarrow$ ☐$+ 5$☐ $\rightarrow n \times n + 5$ or $n^2 + 5$

$1 \rightarrow 6, 2 \rightarrow 9, 3 \rightarrow 14, 4 \rightarrow 21, 5 \rightarrow 30$
so the first five terms of the sequence are 6, 9, 14, 21 and 30.

■ ***Example 4:*** Draw a function machine to show how to evaluate
the expression: (a) $4(5n - 2)$ (b) $4(n + 5) + 3$ (c) $2(n^2 + 5)$

In each case write down the first five terms of the corresponding
sequence:

(a) $n \rightarrow$ ☐$\times 5$☐ $\rightarrow$ ☐$- 2$☐ $\rightarrow$ ☐$\times 4$☐ $\rightarrow 4(5n - 2)$

$1 \rightarrow 12, 2 \rightarrow 32, 3 \rightarrow 52, 4 \rightarrow 72, 5 \rightarrow 92$
so the first five terms are 12, 32, 52, 72 and 92.

(b) $n \rightarrow$ ☐$+ 5$☐ $\rightarrow$ ☐$\times 4$☐ $\rightarrow$ ☐$+ 3$☐ $\rightarrow 4(n + 5) + 3$

$1 \rightarrow 27, 2 \rightarrow 31, 3 \rightarrow 35, 4 \rightarrow 39, 5 \rightarrow 43$
so the first five terms of the sequence are 27, 31, 35, 39 and
43.

(c) $n \rightarrow$ ☐square☐ $\rightarrow$ ☐$+ 5$☐ $\rightarrow$ ☐$\times 2$☐ $\rightarrow 2(n^2 + 5)$

$1 \rightarrow 12, 2 \rightarrow 18, 3 \rightarrow 28, 4 \rightarrow 42, 5 \rightarrow 60$
so the first five terms of the sequence are 12, 18, 28, 42 and
60.

EXERCISE 11

1 Draw a function machine to show how to evaluate the expression:
   (*a*) $4n - 3$ (*b*) $4(n + 3)$ (*c*) $4n^2$
For each write down the first five terms of the corresponding
sequence.

2 Draw a function machine to show how to evaluate the expression:
   (*a*) $4(n - 3) + 5$ (*b*) $4(5n + 3)$ (*c*) $4(n^2 + 7)$
For each write down the first five terms of the corresponding
sequence.

3 Draw a function machine to show how to evaluate the expression:
   (*a*) $5(n + 3) - 2$ (*b*) $3(4n - 2) + 5$ (*c*) $3(n + 1)^2 + 7$
For each write down the first five terms of the corresponding
sequence.

4 Draw a function machine for:
   (*a*) $7(n + 1) + 3$ (*b*) $7n + 10$

For each write down the first five terms of the corresponding sequence.

Are the two sequences the same? Why are the expressions equivalent?

In Question 4 of Exercise 11 the two expressions can be shown to be the same by removing the brackets in

$$7(n + 1) + 3 = 7n + 7 + 3 = 7n + 10.$$

You do not need to simplify more complicated expressions like this before evaluating them, provided you can write down the sequence of operations which enables you to work out the value of the expression for a particular value of n. An example comparing the two methods is shown below.

■ *Example 5:* Evaluate the expressions $2(3n^2 + 5) - 4$ when $n = 7$

(*a*) by drawing a function machine
(*b*) by simplifying the expression by removing the brackets.

(*a*) The function machine is:

$$n \rightarrow \boxed{\text{square}} \rightarrow \boxed{\times 3} \rightarrow \boxed{+ 5} \rightarrow \boxed{\times 2} \rightarrow \boxed{- 4} \rightarrow$$
$$2(3n^2 + 5) - 4$$

with $n = 7$ you get:

$$7 \rightarrow \boxed{\text{square}} \rightarrow 49 \rightarrow \boxed{\times 3} \rightarrow 147 \rightarrow \boxed{+ 5} \rightarrow 152 \rightarrow \boxed{\times 2} \rightarrow$$
$$304 \rightarrow \boxed{- 4} \rightarrow 300$$

(*b*) Simplifying the expression you get:

$$2(3n^2 + 5) - 4 = 6n^2 + 10 - 4 = 6n^2 + 6$$

so when $n = 7$ you have
$6n^2 + 6 = 6 \times 7^2 + 6 = 6 \times 49 + 6 = 294 + 6 = 300$

EXERCISE 12

1   What expression is evaluated by the function machine?

(*a*) $n \rightarrow \boxed{\times 4} \rightarrow \boxed{- 3} \rightarrow \boxed{\times 5} \rightarrow$????

(*b*) $n \rightarrow \boxed{+ 3} \rightarrow \boxed{\times 5} \rightarrow \boxed{- 2} \rightarrow$????

(*c*) $n \rightarrow \boxed{\times 6} \rightarrow \boxed{- 5} \rightarrow \boxed{\times 2} \rightarrow \boxed{- 7} \rightarrow$????

(*d*) $n \rightarrow \boxed{+ 4} \rightarrow \boxed{\times 3} \rightarrow \boxed{- 5} \rightarrow \boxed{\times 2} \rightarrow$????

2 Draw a function machine to show how to evaluate the expression:

(a) $5(4n - 3) + 6$ (b) $4(n + 3)^2 - 5$
(c) $7(4(n + 3) - 2) - 5$

Find the value of each expression when $n = 1$.

3 Evaluate the expression $3(2(n + 1) - 5) + 14$ when $n = 3$

(a) by drawing a function machine
(b) by simplifying the expression by removing the brackets.

Note: You can describe each of the function machines met so far by a rule of the form: $n \rightarrow f(n)$, where $f(n)$ is an expression containing n's.

In Example 5, $n \rightarrow 2(3n^2 + 5) - 4$,

so here $f(n) = 2(3n^2 + 5) - 4$

When $n = 7$, you found that $7 \rightarrow 300$. You can also show this as $f(7) = 300$.

48 FUNCTION NOTATION

Although so far you have used the letter n to represent the position of the nth term in a sequence you could have used any letter to describe the rule. The rule $n \rightarrow 2n + 3$ is the same as the rule $x \rightarrow 2x + 3$. In each case the rule tells you to multiply the number you started with by 2 and then to add 3.

You say that the **image** of n when using this rule is $2n + 3$, whereas the image of x would be $2x + 3$. In either case the image of 5 is $2 \times 5 + 3$ i.e. 13.

Another way of showing the rule $x \rightarrow 2x + 3$ is to use a **mapping** (or arrow) diagram. The rule tells you to *multiply by 2 and then to add* 3.

$$1 \rightarrow 5$$
$$2 \rightarrow 7$$
$$3 \rightarrow 9$$
$$4 \rightarrow 11$$
$$5 \rightarrow 13$$

In general you have $\qquad x \rightarrow 2x + 3.$

The first set of numbers 1, 2, 3, 4, 5 and 6 is called the **domain** of the mapping.

The second set of numbers 5, 7, 9, 11, 13 and 15 is called the **image** or **range** set for this mapping.

In order to distinguish between two different rules a label is often attached. You might use the letter f as the label for the rule $x \to 2x$ and the letter g as the label for the rule $x \to x + 3$. So f represents the *multiply by* 2 rule and g represents the *add* 3 rule.

The notation $f(x)$ is used to represent the image of x when using the f rule. Here $f(x) = 2x$, so $f(5) = 10$ and $f(7) = 14$.

In a similar way the notation $g(x)$ is used to represent the image of x when using the g rule. Here $g(x) = x + 3$, so $g(5) = 8$ and $g(7) = 10$.

Note: the rule $x \to 2x + 3$ is really made up of the two rules f and g. First use the rule f (*multiply by* 2) and then use the rule g (*add* 3) on the result. Using a function machine this would be:

$$x \to \boxed{\text{use } f} \to f(x) \to \boxed{\text{use } g} \to g[f(x)] \text{ or } gf(x)$$

or $\quad x \to \boxed{\times 2} \to 2(x) \to \boxed{+ 3} \to 2x + 3$

You can use the letters gf to represent the composite rule $x \to 2x + 3$.

Note: in this context, gf means use the f rule first and then the g rule. This is not the same as fg which means use the g rule first and then the f rule.

The corresponding function machine for fg is:

$$x \to \boxed{\text{use } g} \to g(x) \to \boxed{\text{use } f} \to f[g(x)] \text{ or } fg(x)$$

or $\quad x \to \boxed{+ 3} \to x + 3 \to \boxed{\times 2} \to 2(x + 3)$

The study of functions and their properties is an important part of mathematics. However it is not intended in this book to cover the work in any formal way, or to do any more than to introduce some basic ideas.

■ *Examples:*

(1) Find the composite rule fg where f represents $x \to 3x$ and g represents $x \to x - 2$. Find also $fg(5)$.

(2) Find the composite rule gf where f represents $x \to x + 3$ and g represents $x \to x^2$. Find also $gf(5)$.

(1) fg means use g (*subtract* 2) first and then use f (*multiply by* 3)

$$x \to \boxed{\text{use } g} \to g(x) \to \boxed{\text{use } f} \to f[g(x)] \text{ or } fg(x)$$

so $x \to \boxed{- 2} \to x - 2 \to \boxed{+ 3} \to 3(x - 2)$

hence fg is $x \to 3(x - 2)$ or $fg(x) = 3(x - 2)$ and
$fg(5) = 3x(5 - 2) = 9$

(2) gf means use f (*add* 3) first and then use g (*square*)

$$x \to \boxed{\text{use } f} \to f(x) \to \boxed{\text{use } g} \to g[f(x)] \text{ or } gf(x)$$

so $x \to \boxed{+3} \to x + 3 \to \boxed{\text{square}} \to (x + 3)^2$

hence gf is $x \to (x + 3)^2$ or $gf(x) = (x + 3)^2$ and
$gf(5) = (5 + 3)^2 = 64$

EXERCISE 13

1 Write down in the form of $x \to \ldots$ the rules:

    (*a*) *double and then add* 7.   If this rule is f find $f(3)$
    (*b*) *add* 7 *and then double.*   If this rule is g find $g(3)$
    (*c*) *add* 5 *and then square.*   If this rule is h find $h(3)$
    (*d*) *square and then add* 5.   If this rule is k find $k(3)$

2 If f is the rule *subtract* 3 and g is the rule *multiply by* 2 find:

    (*a*) $f(5)$ (*b*) $g(9)$ (*c*) the rule fg (*d*) the rule gf

3 Using the rules f, g, h and k from Question 1, find:

    (*a*) $gf(3)$ (*b*) $fg(3)$ (*c*) $hk(3)$ (*d*) $kh(4)$
    (*e*) the rule kf

The opposite (or **inverse**) of the rule *add* 3 is the rule *subtract* 3. If f is the rule $x \to x + 4$, then its inverse is the rule $x \to x - 4$. In the same way if g is the rule $x \to 3x$, its inverse is $x \to x \div 3$. The inverse of the rule f is called f^{-1}. The inverse of g is called g^{-1}.

Note: if $f(7) = 11$, then $f^{-1}(11) = 7$, also if $g(7) = 21$, then $g^{-1}(21) = 7$.

■ *Example 1:* Find the inverse of the rule $x \to 3x + 4$.

This rule tells you first to *multiply by* 3 and then to *add* 4. To find the inverse of this rule you need to 'undo' the *add* 4 first and then to 'undo' the *multiply by* 3, i.e. you first *subtract* 4 and then *divide by* 3. So the inverse of $x \to 3x + 4$ is $x \to (x - 4) \div 3$.

    This process can be seen most easily using a function machine.
    The function machine for the rule $x \to 3x + 4$ is

$$x \to \boxed{\times 3} \to 3x \to \boxed{+4} \to 3x + 4$$

If you want to undo this process you must run the machine 'backwards'.

i.e. $x \leftarrow \boxed{\div 3} \leftarrow 3x \leftarrow \boxed{-4} \leftarrow 3x + 4$

So reading from right to left you first *subtract* 4 and then *divide by* 3. This is the rule $x \to (x - 4) \div 3$.

■ *Example 2:* Find the inverse of the rule $x \to 3(x + 4) - 2$

The function machine for the rule $x \to 3(x + 4) - 2$ is

$$x \to \boxed{+4} \to \boxed{\times 3} \to \boxed{-2} \to 3(x + 4) - 2$$

If you run the machine 'backwards' you get

$$x \leftarrow \boxed{-4} \leftarrow \boxed{\div 3} \leftarrow \boxed{+2} \leftarrow 3(x + 4) - 2$$

So reading from right to left you first *add* 2, then you *divide by* 3 and then you *subtract* 4. This gives the rule $x \to (x + 2) \div 3 - 4$.

You can check that this is correct by trying out a specific value. Say $x = 6$.

Using the rule:

$$x \to 3(x + 4) - 2, \quad 6 \to 3 \times (6 + 4) - 2 = 28$$

Using the inverse rule

$$x \to (x + 2) \div 3 - 4, \quad 28 \to (28 + 2) \div 3 - 4 = 6$$

■ *Example 3:* Find the number which gives 19 as the result when using the rule $x \to 3x + 4$

The function machine for the rule $x \to 3x + 4$ is:

$$x \to \boxed{\times 3} \to \boxed{+4} \to 3x + 4$$

You know that the result has to be 19, so you must run the machine 'backwards' to find the original number. In this case it is 5.

i.e. $5 \leftarrow \boxed{\div 3} \leftarrow 15 \leftarrow \boxed{-4} \leftarrow 19$

EXERCISE 14

1 (a) Draw a function machine for the rule $x \to 4x + 3$
        What is the result if the number used is 5?
   (b) Now run the machine backwards to find the inverse of the rule.
        Check that 5 is the result if the number used is 23.

2 (*a*) Draw a function machine for the rule $x \to 5(x - 4) + 3$
What is the result if the number used is 8?

(*b*) Now run the machine backwards to find the inverse of the rule.
Check that 8 is the result if the number used is 23.

3 Find the inverse of the rule $x \to 2x - 3$
Use this to find what number gives 11 when using the original rule.

4 Find the inverse of the rule $x \to 5(2x - 3) + 7$
Hence find the value of x for which $5(2x - 3) + 7 = 22$

49 AN INTRODUCTION TO SOLVING EQUATIONS

The ideas introduced in *Example 3* on page 57 enable you to solve simple problems of the form: For what value of x is $3x + 4 = 19$? This is an example of a linear equation which you will consider in greater detail in the next chapter.

■ *Example 1:* For what value of x is $5x - 7 = 8$?

First draw the function machine which gives the expression on the left-hand side of the equation.

$$x \to \boxed{\times 5} \to \boxed{- 7} \to 5x - 7$$

You want to find the number which gives 8 when using this machine. To do this draw the inverse machine using 8 as the starting number.

$$3 \leftarrow \boxed{\div 5} \leftarrow 15 \leftarrow \boxed{+ 7} \leftarrow 8$$

This shows that $5x - 7 = 8$ when $x = 3$. To check substitute $5 \times 3 - 7 = 8$.

■ *Example 2:* For what value of x is $5(x - 3) = 20$?

First draw the function machine for the left-hand side of the equation.

$$x \to \boxed{- 3} \to \boxed{\times 5} \to 5(x - 3)$$

Now draw the inverse machine and use the right-hand number i.e. 20

$$7 \leftarrow \boxed{+ 3} \leftarrow 4 \leftarrow \boxed{\div 5} \leftarrow 20$$

So $5(x - 3) = 20$ when $x = 7$. Check: $5(7 - 3) = 5 \times 4 = 20$

■ *Example 3:* For what value of n is $3(2n - 5) + 4 = 55$?

First draw the function machine for the left-hand side of the equation.

$$n \rightarrow \boxed{\times 2} \rightarrow \boxed{-5} \rightarrow \boxed{\times 3} \rightarrow \boxed{+4} \rightarrow 3(2n - 5) + 4$$

Now draw the inverse machine and use the right-hand number i.e. 55

$$11 \leftarrow \boxed{\div 2} \leftarrow 22 \leftarrow \boxed{+5} \leftarrow 17 \leftarrow \boxed{\div 3} \leftarrow 51 \leftarrow \boxed{-4} \leftarrow 55$$

So $3(2n - 5) + 4 = 55$ when $n = 11$.
Check: $3(2 \times 11 - 5) + 4 = 3 \times 17 + 4 = 55$

Providing you understand how the expression on the left-hand side of the equation is built up and you can draw the corresponding function machine, this type of problem is then a matter of working backwards using the inverse machine. Remember to check your answer by substitution.

EXERCISE 15

1 For what value of x is:
   (*a*) $3x + 7 = 19$ (*b*) $5(x - 4) = 30$ (*c*) $x \div 5 + 3 = 8$.

2 For what value of x is:
   (*a*) $7(x + 4) - 3 = 67$ (*b*) $2(3x - 7) + 9 = 25$.

3 For what value of n is:
   (*a*) $3(4n - 7) + 5 = 44$ (*b*) $2(5(n + 1) - 3) + 7 = 41$.

6

Linear Equations

50 MEANING OF AN EQUATION

If you know that 5 times a certain number is 40, a simple process in arithmetic enables you to calculate that the number is 8.

In algebraic form the problem could be expressed as follows:

Let n = the unknown number. Then the question can be put in this way:

If $5n = 40$, what is the value of n?

The statement $5n = 40$ is called an **equation**. It is a statement of equality, but it also implies that a value of n is required which will make the left-hand side of the equation equal to the right, or which 'satisfies the equation'. The process of finding the value of n which thus satisfies the equation is called **solving the equation**.

The solution of the above equation involves no more than the division of the right-hand side by the coefficient of n, and could be stated thus:

$$5n = 40$$
$$\therefore \quad n = 40 \div 5$$
$$\therefore \quad n = 8.$$

The solution of an equation is rarely so simple as this. Equations usually consist of more or less complicated expressions on both sides of the equation. By various operations you aim to reduce the equation to the simple form above. You can then easily find the value of the unknown letter. These operations will be illustrated in the examples which follow.

51 SOLVING AN EQUATION

■ *Example 1:* If 8 times a number is decreased by 5 the result is 123. What is the number?

This simple problem could be solved mentally, but it will serve as an introduction to the process of solution.

Let n = the number.

Then $8n - 5$ is the expression which states algebraically '8 times the number decreased by 5'.

But this is equal to 123.

Hence you can form the equation

$$8n - 5 = 123.$$

This is the first step that must always be taken – to formulate the equation. Then you proceed to the solution – i.e., to find the value of n which satisfies it.

Now, the above statement means that 123 is 5 less than 8 times the number, or, if 123 is increased by 5, it is equal to 8 times the number.

∴ you can write the equation in this form

$$8n = 123 + 5.$$

Thus you have almost reached the form you want, after which you can find the solution.

This step was reached in effect, by transferring the 5 to the right-hand side, leaving only a multiple of n, the unknown number, on the left side. In this transference the argument involved changing the sign of the 5.

The same result could be obtained as follows:

Since $8n - 5 = 123,$

if each side is increased by 5 the result will be that you will be left with $8n$ only on the left-hand side and the two sides will still be equal. You will have as your equation:

$$8n - 5 + 5 = 123 + 5$$
whence $$8n = 128$$
and $$n = 16.$$

This device is employed in the solution of practically every equation and it depends on the fact that:

(A) *If the same number is added to, or subtracted from, both sides of an equation, the two sides will again be equal.*

As a working rule this is equivalent to transferring a number from one side of an equation to the other at the same time changing its sign, i.e., change + to − and − to +.

A principle similar to the above which will be employed later, is:

(B) *If both sides of an equation are multiplied or divided by the same number, the two sides of the new equation will be equal. If the multiplier is −1 or any negative number, both sides change signs.*

■ *Example 2:* In Section 4 you saw that three consecutive odd numbers could be expressed algebraically by

$$2n + 1, 2n + 3, 2n + 5$$

where *n* is any integer.

Now suppose that you wish to solve this problem:

The sum of three consecutive odd numbers is 81. What are the numbers?

As stated above, the first step is to form an equation. This usually means putting into algebraic form the facts which are given about the unknown number or numbers.

First, as above, represent the three odd numbers by

$$2n + 1, 2n + 3, 2n + 5.$$

Then express algebraically the fact that their sum is 81 by writing the equation

$$(2n + 1) + (2n + 3) + (2n + 5) = 81.$$

The use of the brackets helps to make the statement clear.

Now remove the brackets and get:

$$2n + 1 + 2n + 3 + 2n + 5 = 81.$$

Adding like terms $\qquad 6n + 9 = 81$

whence, as above $\qquad 6n = 81 - 9$

or $\qquad 6n = 72.$

$\therefore \qquad n = 12.$

You can now obtain the odd numbers by substitution of $n = 12$ in $2n + 1, 2n + 3, 2n + 5$.

$\therefore$ the numbers are 25, 27, 29.

You should check this by ascertaining that their sum is 81.

52 WORKED EXAMPLES

Equations arise out of practical problems in a variety of ways, and examples will be given later, but it is desirable that you should first have sufficient practice in the methods of solving equations. Examples of equations will therefore be worked out and provided for practice which will have no relation to any special problems.

It is common in such practice equations to use letters at the end of the alphabet, x, y and z, to represent the unknown numbers, and, when necessary, letters at the beginning of the alphabet, a, b, c, etc., to represent known numbers.

This choice of letters is due to Descartes (seventeenth century).

■ *Example 1:* Solve the equation:

$$6x - 5 = 2x + 9.$$

The general plan adopted is to collect the terms involving the unknown number, x, on the left side, and the other terms on the right.

Transferring the x term from the right side you get:

$$6x - 2x - 5 = 9.$$

Transferring the -5,

$$6x - 2x = 9 + 5$$
$$\therefore \qquad 4x = 14$$
and $$x = \frac{7}{2}.$$

Note: With practice the two transference steps could be taken together.

Check. You can always check the accuracy of the solution to an equation by substituting the value found in both sides of the original equation. In the above case:

Left side $\left(6 \times \dfrac{7}{2}\right) - 5 \ = 16.$

Right side $\left(2 \times \dfrac{7}{2}\right) + 9 = 16.$

The two sides are equal.

$\therefore \qquad\qquad x = \dfrac{7}{2}$ satisfies the equation.

■ *Example 2:* Solve the equation:

$$10(x - 4) = 4(2x - 1) + 5.$$

First simplify both sides by removing brackets.

Then $10x - 40 = 8x - 4 + 5.$

Transferring $8x$ to the left side and -40 to the right,

$$10x - 8x = 40 - 4 + 5.$$
∴ $2x = 41$
and $x = 20\tfrac{1}{2}.$

Check.

Left side $10(20\tfrac{1}{2} - 4) = 10 \times 16\tfrac{1}{2} = 165.$
Right side $4(2 \times 20\tfrac{1}{2} - 1) + 5 = 160 + 5 = 165.$
∴ $x = 20\tfrac{1}{2}$ satisfies the equation.

■ *Example 3:* Solve the equation:

$$\frac{3x}{5} + \frac{x}{2} = \frac{5x}{4} - 3.$$

When the equation involves fractions, the first step, in general, towards simplification is to 'clear the fractions'. This is effected by multiplying throughout by such a number that the fractions disappear. This is justified by Principle B, Section 51.

The smallest number which will thus clear the fractions is the L.C.M. of their denominators, in this case 20.

Multiplying every term on both sides by 20 you get:

$$\left(\frac{3x}{5} \times 20\right) + \left(\frac{x}{2} \times 20\right) = \left(\frac{5x}{4} \times 20\right) - (3 \times 20).$$

∴ $12x + 10x = 25x - 60.$
∴ $12x + 10x - 25x = -60$
                   $-3x = -60.$
∴ $x = -60 \div -3.$
∴ $x = 20$ (by the law of signs).

You should check this solution as in the previous examples.

■ *Example 4:* Solve the equation:

$$4x - \frac{x - 2}{3} = 5 + \frac{2x + 1}{4}.$$

Multiplying throughout by 12

$$48x - 4(x - 2) = 60 + 3(2x + 1). \quad \text{(Section 30)}$$

Clearing brackets

$$48x - 4x + 8 = 60 + 6x + 3.$$
$$\therefore \quad 48x - 4x - 6x = 60 + 3 - 8.$$
$$\therefore \quad 38x = 55$$

and

$$x = \frac{55}{38}.$$

Check by substitution.

<u>EXERCISE 16</u>

Solve the following equations:

1 (a) $14x = 35$ (b) $1.5x = 30$
2 (a) $\frac{3}{2}x = 24$ (b) $\frac{5}{8}x = 50$
3 (a) $3x = -48$ (b) $-2x = 20$
4 (a) $-\frac{1}{2}x = -40$ (b) $-\frac{1}{3}x = 24$

5 (a) $\dfrac{x}{0.6} = 21$ (b) $\dfrac{2x}{0.2} = -0.8$

6 (a) $5x + 8 = 24$ (b) $10x - 9 = 41$
7 (a) $3x - 4.7 = 2.8$ (b) $2.5x + 50 = 80$
8 (a) $3x - 5 = 2x + 3$ (b) $6y + 11 = 3y + 15$
9 (a) $2x + 6 = 14 - 3x$ (b) $z + 20 = 5z - 44$
10 (a) $3y - 1.5 = 7y - 8.7$ (b) $4.8x + 52 = 3.2x - 20$

11 (a) $3y - \dfrac{44}{5} = 2y + 6$ (b) $x + \dfrac{x}{15} = 20$

12 (a) $\dfrac{3x}{5} - \dfrac{2}{3} = 0$ (b) $\dfrac{x}{0.1} + \dfrac{x}{0.2} = 5$

13 $4(2x - 5) = 3(2x + 8)$
14 $3x - 2(x + 4) = 5x - 28$
15 $2(x + 5) - 3(x - 6) = 20$
16 $5(y - 1) - 2(y + 6) = 2y + 12$
17 $2(x - 1) + 3(x + 4) = 4(x + 1) - (x - 5)$
18 $3(x - 7) - 3(2x - 4) = 4(x + 3)$
19 $\frac{1}{2}(3y + 6) - \frac{1}{3}(2y - 4) = 20$

20 (a) $\dfrac{x}{8} - \dfrac{4}{5} = 0$ (b) $\dfrac{2x}{5} - 3 = 8$

21 (a) $\dfrac{7x}{8} - (x - 2) = 12$ (b) $\dfrac{3x}{2} - \dfrac{x}{3} = \dfrac{5(x - 4)}{6}$.

22 $\dfrac{2x - 5}{3} - \dfrac{3x - 1}{4} = \dfrac{3}{2}$.

23 $\dfrac{7x - 6}{4} - \dfrac{5x + 3}{7} - \dfrac{6x - 1}{14} = 3$

24 $\frac{1}{2}(x + 3) - \frac{2}{5}(x - 3) = x + 9$

25 $12(5 - x) - 3(3x - 4) = 23$

26 $\frac{3}{2}(x - 1) - \frac{1}{3}(5 - x) = 2\frac{1}{3} - 3(x - 3)$

27 $\dfrac{y}{7} - \dfrac{1 - y}{5} = 2 - \dfrac{3y}{5}$

28 $\dfrac{2x + 4}{1.5} - \dfrac{3x - 5}{2.5} = 1$

29 Solve for n the equation $2n = 0.58(12 - n)$.

30 For what value of r is 18.4 equal to $2(3.5r - 1)$?

31 Find x when $\dfrac{7.5}{x} = \dfrac{5}{2}$.

32 Find c if $\dfrac{18}{2c} = 3.8$.

33 If $C = \dfrac{V}{R}$, find V when $C = 8$, $R = 4.5$.

34 For what value of x is $3(x - 5)$ equal to $\dfrac{4x + 3}{2}$?

53 PROBLEMS LEADING TO SIMPLE EQUATIONS

The methods of solving problems by means of simple equations are illustrated by the following examples. Here is the general method of procedure.

(1) Having decided which is the unknown quantity, represent it by a letter, such as x, stating clearly the units employed when necessary.

(2) Form an equation which represents the facts provided by the problem about the unknown quantity.

(3) Solve the equation.

■ *Example 1:* At a school fête there were 10 more children than adults. The adults paid 20p each, the children 15p each, and the total receipts were £16.20. How many adults and children were there at the fête?

There are two unknown quantities: the number of adults and the number of children. But if the number of adults is known, the number of children is 10 more.

∴   Let x = the number of adults.
Then $x + 10$ = the number of children.
The facts supplied are represented as follows:

$$20x = \text{the money paid by the adults, in pence;}$$
$$15(x + 10) = \text{the money paid by the children, in pence;}$$
$$1620 = \text{the total amount paid, in pence.}$$

The equation which connects these is consequently:

$$20x + 15(x + 10) = 1620.$$
∴ $\qquad 20x + 15x + 150 = 1620$
$$35x = 1620 - 150$$
$$35x = 1470.$$
∴ $\qquad\qquad\qquad x = 42.$
∴   the number of children is $x + 10 = 52$.
∴   the solution is 42 adults, 52 children.

■ *Example 2:* You travel in a car from town A to town B at an average speed of 64 km/h. On your return journey your average speed is 80 km/h. You take 9 hours for the double journey (not including stops). How far is it from A to B?

The unknown quantity is the distance from A to B.
Let x = distance in kilometres.
Now distance = speed × time.

∴ $$\frac{\text{distance}}{\text{speed}} = \text{time.}$$

Then $\qquad$ time for 1st journey = $\dfrac{x}{64}$ hours.

Then $\qquad$ time for 2nd journey = $\dfrac{x}{80}$ hours.
But the total time is 9 hours.
∴ the equation is

$$\frac{x}{64} + \frac{x}{80} = 9.$$

To clear the fractions multiply throughout by 320.

$$\therefore \quad \left(\frac{x}{64} \times 320\right) + \left(\frac{x}{80} \times 320\right) = 9 \times 320$$

$$5x + 4x = 9 \times 320$$
$$\therefore \quad 9x = 9 \times 320$$
and $\quad x = 320 \text{ km.}$

You should check this solution against the information in the question.

■ *Example 3:* The hourly wages of two people A and B engaged on the same kind of work were £4.80 and £3.70 respectively. It was agreed to increase the two wages by the same amount so that A's wage was $\frac{6}{5}$ of B's wage. What increase was given?

The unknown quantity is the money to be added to the wages.
Let x = the amount of increase in pence.

Then $\quad (480 + x)$ pence = A's new wages.
$\quad (370 + x)$ pence = B's new wages.
Then by the data

$$480 + x = \tfrac{6}{5}(370 + x).$$

Clearing fractions

$$5(480 + x) = 6(370 + x).$$
$$\therefore \quad 2400 + 5x = 2220 + 6x.$$
$$\therefore \quad 2400 - 2220 = 6x - 5x$$
and $\quad 180 = x.$

∴ the hourly increase is 180p or £1.80.

You should check this by adding it to each of the hourly wages and ascertaining if one is $\frac{6}{5}$ of the other.

EXERCISE 17

1 From three times a certain number, n, 6 is subtracted. The result is equal to twice the number with 6 added. What is the value of n?

2 There is a number such that when it is multiplied by 5 and then 14 is subtracted, the result is 348.5. Find the number.

3 From 5 times a certain number, 189 is subtracted, and the remainder is one half the original number. What is that number?

4 One-third of a number added to four-fifths of itself is equal to 17. What is the number?

5 When 9 is subtracted from 6 times a certain number, the result is 45 more than twice the number. Find the number.

6 The sum of three consecutive odd numbers is 69. What are the numbers?

7 You walk from one town to another at an average speed of 2.5 km/h. On the return you quicken your average speed to 3 km/h. The time taken for the double journey was 7 h 20 min. How far are the two towns apart?

8 The sum of a number and 4 per cent of itself is 41.6. What is the number?

9 The perimeter of a rectangle is 44 cm. If one of the two adjacent sides is 1.8 cm longer than the other, what are the lengths of the sides?

10 Some people agree to pay equally for the use of a boat, and each pays 15p. If there had been two more people in the party, each would have paid 10p. How many people were there and how much was the hire of the boat?

11 A teacher distributes £2 among 20 children, giving 5p each to some and 25p to the rest. How many children received 25p each?

12 A man is four times as old as his son. In four years' time he will be three times as old. What are their ages now?

13 The connection between the degrees on the Celsius and

Fahrenheit thermometers is that $x°F = \left(\dfrac{9x}{5} + 32 \right)°C.$

What number of degrees Celsius is equivalent to 86°F?

14 A bookseller buys 120 volumes of a certain series of books. Some are sold at the published price of £18 each and the rest are sold in a sale at £12 each. If the total receipts are £1,920, find how many volumes were sold at each price.

15 A bus is carrying 32 passengers, some with 60p tickets and the remainder with £1 tickets. If the total receipts from these passengers are £22.80, find the number of 60p fares.

7

Formulae

54 PRACTICAL IMPORTANCE OF FORMULAE

One of the most important applications of elementary algebra is in the use of formulae. In every form of applied science and mathematics, such as mechanical engineering, electrical engineering, aeroplane construction, etc., formulae are constantly employed, and their interpretation and manipulation are essential.

55 TREATMENT OF FORMULAE

In the first chapter of this book some very easy examples of formulae were introduced. With the assistance of a greater knowledge of algebraical symbols and operations and with increasing skill in their use, you can now proceed to more difficult types.

Formulae involve three operations:

(1) construction;   (2) manipulation;   (3) evaluation.

The construction of formulae cannot be indicated by any specified rules or methods. A knowledge of the principles of algebra and skill in their application are necessary. But in general you are concerned with formulae which have already been evolved. What you need is skill in using them; and as your knowledge of algebra increases, so you will be better equipped for dealing with new examples.

The manipulation and evaluation of formulae are closely associated. A formula may need to be re-arranged or simplified before any substitution of values may be made. Experience alone will guide you as to what manipulation is desirable in order to reach a form which is the most suitable for evaluation or some other purpose. Clear arrangement of working is always essential for accuracy.

56 WORKED EXAMPLES

■ *Example 1:* Find a formula for the total area (A) of the surface of a square pyramid as in Fig. 13 when $AB = a$ and $OQ = d$ (remember that the area of a triangle is $\frac{1}{2}$ the base $\times$ the height).

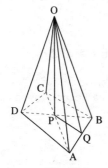

Fig. 13

OQ is perpendicular to AB, and represents the height of the $\triangle AOB$.

The total area is made up of:

(1) area of base;
(2) areas of the four $\triangle$s, of which AOB is one.

Area of base $= a^2$.
Area of each $\triangle = \frac{1}{2}ad$.
∴ area of all $\triangle$s $= 4 \times \frac{1}{2}ad = 2ad$.
∴ total area of pyramid $= a^2 + 2ad$
or $\qquad A = a(a + 2d)$.

■ *Example 2:* If $L = \dfrac{W(T - t)}{w} - t$,

find L when $t = 8.5$, $w = 115$, $W = 380$, $T = 28.5$.

Substituting in $\quad L = \dfrac{W(T - t)}{w} - t$.

$$L = \frac{380(28.5 - 8.5)}{115} - 8.5$$

$$= \frac{380 \times 20}{115} - 8.5$$

$$= 66.1 - 8.5 \text{ (approx.)}$$
$$= 57.6 \text{ to 1 decimal place (approx.)}$$

EXERCISE 18

1 If $s = ut + \frac{1}{2}at^2$, find s when $u = 15$, $t = 5$, $a = 8$.

2 The volume of a cone, V, is given by the formula $V = \frac{1}{3}\pi r^2 h$, where r = radius of base, h = height of cone.

Find V when $r = 3.5$, $h = 12$, $\pi = \frac{22}{7}$.

3 The volume of a sphere is given by the formula $V = \frac{4}{3}\pi r^3$, where r = radius. Find V when $r = 3$, $\pi = \frac{22}{7}$.

4 If $E = \frac{Wv^2}{2g}$, find E when $W = 15.5$, $v = 18.8$, and $g = 32$.

5 From the formula $C = \frac{E + e}{R + r}$, find C when $E = 17.6$, $e = 1.5$,

$R = 28.4$, $r = 2.6$.

6 In a suspension bridge the length of the cable employed is given by the formula $L = l + \frac{8d^2}{3l}$, where

L = length of the cable,
d = dip of the centre of the cable,
l = length of the span of the bridge,

all measurements being in metres. Find L when $d = 6$, $l = 56$.

7 A formula for the loading of beams is $W = \frac{4kbd^2}{l}$.

Find W when $k = 45$, $b = 2$, $d = \frac{1}{2}$, $l = 20$.

8 If $s = \frac{n(n + 1)(2n + 1)}{6}$, find s when $n = 8$.

9 The following formula is used in connection with pile driving, $L = \frac{Wh}{d(W + P)}$. Find L when $W = 5$, $d = 1.5$, $P = 19$, $h = 4.5$.

10 If $R = W(x + 3t)$, find R when $W = 210$, $x = 6.5$, $t = 0.04$.

11 From the formula $H = \frac{plAN}{33\,000}$ find H when $p = 18$, $l = 2$,

$A = 80$, $N = 360$.

57 TRANSFORMATION OF FORMULAE

In the formulae which have been examined you can see that one quantity is expressed in terms of other quantities and the formula expresses the relations between them. Thus in the formula for the volume of a cone ($V = \frac{1}{3}\pi r^2 h$, Exercise 18, No. 2) this volume is expressed in terms of the height and the radius of the base.

But you may need to express the height of the cone in terms of the volume and the radius of the base. In that case you would write the formula in the form:

$$h = \frac{3V}{\pi r^2},$$

that is, the formula has been transformed.

When one quantity is expressed in terms of others, as in $V = \frac{1}{3}\pi r^2 h$, the quantity thus expressed, in this case V, is sometimes called the **subject of the formula**.

When the formula was transformed into

$$h = \frac{3V}{\pi r^2},$$

the subject of the formula is now h. This process of transformation is called **changing the subject of the formula**.

The transformation of formulae often requires skill and experience in algebraical manipulation; the following examples will help to illustrate the methods to be followed.

58 WORKED EXAMPLES

■ *Example 1:* From the formula $T = \dfrac{\pi f d^3}{16}$, find:

(1) f is terms of the other quantities.
(2) d in terms of the other quantities.

From
$$T = \frac{\pi f d^3}{16},$$

clearing the fraction $16T = \pi f d^3$ (1)
or $\pi f d^3 = 16T,$

dividing throughout by πd^3 $f = \dfrac{16T}{\pi d^3}.$

From (1) dividing throughout by πf

$$d^3 = \frac{16T}{\pi f}.$$

$\therefore$
$$d = \sqrt[3]{\frac{16T}{\pi f}}.$$

■ *Example 2:* Transform the formula
$$L = l + \frac{8d^2}{3l}$$

into one which expresses d in terms of the other quantities.

$$L = l + \frac{8d^2}{3l}.$$

Clearing fractions: $\quad 3lL = 3l^2 + 8d^2$

or $\qquad\qquad\qquad 8d^2 = 3lL - 3l^2$.

$\therefore \qquad\qquad\qquad d^2 = \frac{3lL - 3l^2}{8}.$

$\therefore \qquad\qquad\qquad d = \pm\sqrt{\frac{3lL - 3l^2}{8}}.$

■ *Example 3:* The speed, V, of water flowing through a pipe, occurs in the formula,

$$h = 0.03 \, \frac{L}{D} \times \frac{V^2}{2g}.$$

Change the subject of the formula to V.

Write the formula as
$$0.03 \, \frac{L}{D} \times \frac{V^2}{2g} = h.$$

Divide both sides by $0.03 \, \dfrac{L}{D}$,

$$\frac{V^2}{2g} = h \div \frac{0.03L}{D}$$

$$= h \times \frac{D}{0.03L}.$$

$$= \frac{hD}{0.03L} .$$

$$\therefore \qquad V^2 = \frac{2ghD}{0.03L} .$$

$$\therefore \qquad V = +\sqrt{\frac{2ghD}{0.03L}} .$$

■ *Example 4:* If $a - b = x(c - nd)$ find n in terms of the other letters.

Fix your attention on the term containing n, namely nd. Try to isolate this term from the other terms.

∴ divide both sides by x.

Then
$$\frac{a - b}{x} = c - nd.$$

Transferring c

$$\frac{a - b}{x} - c = -nd.$$

$$\therefore \qquad nd = c - \frac{a - b}{x} ,$$

$$\therefore \qquad n = \frac{1}{d}\left\{ c - \frac{a - b}{x} \right\} .$$

■ *Example 5:* The period of oscillation of a simple pendulum is given by the formula

$$t = 2\pi\sqrt{\frac{l}{g}} .$$

Find l in terms of the other quantities.

$$t = 2\pi\sqrt{\frac{l}{g}} .$$

Square both sides.

Then
$$t^2 = 4\pi^2 \times \frac{l}{g} ,$$

$$\therefore \qquad t^2 g = 4\pi^2 l$$

or
$$4\pi^2 l = gt^2.$$

∴
$$l = \frac{gt^2}{4\pi^2}.$$

EXERCISE 19

In each example where a square root ($\sqrt{}$) is involved you should clarify which sign ($+$ or $-$) should be taken.

1 The formula for the area (A) of a circle, in terms of its radius (r) is $A = \pi r^2$. Change the subject of the formula to express r in terms of the area.

2 Transform the formula for the volume of a sphere – namely $V = \frac{4}{3}\pi r^3$ (see Exercise 18, No. 3) – into a formula in which r is expressed in terms of the volume.

3 Change the formula for the volume of a cone – namely, $V = \frac{1}{3}\pi r^2 h$ – to a formula in which the subject is r.

4 The horse power of a motor is given by the formula

$H = \dfrac{EC}{825}$. Express this as a formula for C.

5 The lifting force of an electro-magnet is given by the formula

$F = \dfrac{B^2 A}{112 \times 10^5}$, where F is the force.

Transform this into a formula of which the subject is B.

6 The amount of sag, d, in a beam under certain conditions is

given by the formula $d = \dfrac{Wl^3}{48EI}$.

Express this as a formula expressing l in terms of the other quantities.

7 If $v^2 = u^2 + 2as$, express s in terms of u, v, and a. Find the value of s when $u = 15$, $v = 20$, and $a = 5$.

8 There is an electrical formula $I = \dfrac{V}{R}$.

Express this (1) as a formula for V and (2) as a formula for R. Find I if $V = 2$ and $R = 20$.

9 If $n^2 r + 1 = NR$, rearrange the expression so that it becomes a formula for n. Find the value of n when $N = 25$, $R = 2$, $r = 0.81$.

10 The relation of the volume (v) of a mass of gas to the pressure (p) on it is given by the law $pv = k$.

In a certain experiment when $p = 84$, $v = 12$. Find the value of k and then express the formula giving v in terms of p and the value of k.

59 LITERAL EQUATIONS

The operations employed in changing the subject of a formula are the same in principle as those used in the solution of equations. One essential difference from the equations dealt with in Chapter 6 is that whereas these were concerned with obtaining numerical values when solving the equations in the formula the quantity which is the subject of the formula is expressed in terms of other quantities, and its numerical value is not determined, except when the numerical values of these quantities are known.

It is frequently necessary, however, to solve equations in which the values of the unknown quantities will be found in terms of letters which occur in the equation. Such equations are termed **literal equations**. The methods of solution are the same in principle as those employed in Chapter 6. They are illustrated in the following examples.

60 WORKED EXAMPLES

■ *Example 1:* Solve the equation $5x - a = 2x - b$.

As pointed out previously (Section 52), x is understood as standing for the unknown quantity and the use of the letters a and b marks the difference between this kind of equation and those of Chapter 6. The methods by which the equation is solved are the same, however, a and b being treated in the same way as ordinary numerals.

$$5x - a = 2x - b$$

Transferring $2x$ and $-a$ respectively from one side to the other, and changing the signs in so doing,

$$5x - 2x = a - b$$
$$\therefore \qquad 3x = a - b$$
$$\text{and} \qquad x = \frac{a - b}{3}.$$

■ *Example 2:* Solve for x

$$a(x - 2) = 5x - (a + b).$$

Removing brackets
$$ax - 2a = 5x - a - b.$$

Transferring $ax - 5x = 2a - a - b$
or $ax - 5x = a - b.$

This introduces a point of difference from numerical equations. With the latter you add the terms involving x by adding their coefficients. The addition in this case cannot, however, be made arithmetically. Algebraically the sum of the coefficients of x is

$$a + (-5) \text{ or } a - 5.$$

$\therefore$ you write $(a - 5)x = a - b.$

Dividing both sides by the coefficient of x

$$x = \frac{a - b}{a - 5}.$$

EXERCISE 20

Solve the following equations for x assuming all constants are non-zero:

1 $5x - 4a = 0$

2 $5x - 3a = 7a$

3 $8x - p = 3x + 4p$

4 $3x + 2b = 2(x + 3b)$

5 $ax + b = 3a - b$

6 $b(x - p) = c$

7 $2a - b = b - bx$

8 $3(ax - 2) + 25b = 6b$

9 $p(x - q) = x(p - q)$

10 $\dfrac{ax}{2} - 4 = \dfrac{b}{3} - 6.$

11 $ax - 4b = bx - b$

12 $\dfrac{x}{a} + \dfrac{x}{3a} = 4 - x.$

13 $a(x - a) = b(x + b)$

14 $\dfrac{x}{m} - \dfrac{1}{n} = \dfrac{3x}{mn}.$

8

Simultaneous Equations

61 SIMPLE EQUATIONS WITH TWO UNKNOWN QUANTITIES

The equations considered in Chapter 6 contained only one unknown quantity whose value it was required to determine. But many of the formulae quoted in Chapter 7 contain several quantities. Cases may therefore occur in which it will be required to find the value of more than one of these. Similar problems arise which, for their solution, involve the determination of more than one unknown.

62 SOLUTION OF SIMULTANEOUS EQUATIONS

A simple problem will serve to illustrate the above statement. Suppose you are told

The sum of two numbers is 10: what are the numbers?

Let the two numbers be represented by x and y.
Then you know that

$$x + y = 10.$$

It is evident that there is an infinite number of solutions of this equation, such as $(1, 9)$, $(2, 8)$, $(3.5, 6.5)$, etc.

If the equation be written in the form:

$$x = 10 - y.$$

This gives x in terms of y.

Whatever value you give to y in this equation, you can find a corresponding value of x, and each pair of values gives a solution of the equation.

If a second condition has to be satisfied you can determine which of these pairs works.

For example, suppose that, in addition to the statement that the sum of the numbers is 10, you are also told that one of them is four times the other; then there is only one set of the pairs of values referred to above which will satisfy both the conditions.

For if $$x + y = 10 \qquad (1)$$
and $$x = 4y \qquad (2)$$

substituting for x in equation (1) you get:

$$4y + y = 10$$
$$5y = 10$$
$$y = 2,$$

and since $$x = 4y,$$
then $$x = 8.$$

∴ the solution which satisfies both equations simultaneously is $x = 8$, $y = 2$, and clearly there is no other solution.

For this reason such equations are called **simultaneous equations.**

It is clear that if there are two unknown quantities whose values are required uniquely, it must be possible to form two separate equations connecting them.

The methods employed in solving these equations are shown in the following examples.

63 Worked examples

■ *Example 1:* Solve the equations:

$$2x + y = 21 \qquad (1)$$
$$3x + 4y = 44 \qquad (2)$$

In the method employed in this example you begin by obtaining one letter in terms of the other. The more convenient one is chosen, and in this case from equation (1) $2x + y = 21$, you get:

$$y = 21 - 2x \qquad (3)$$

You could have found x in terms of y, but this would involve fractions and is not so convenient although it ultimately will give the same answer.

Substituting in equation (2) the value of y thus obtained from equation (1)

$$3x + 4(21 - 2x) = 44.$$

Thus you reach a simple equation with one unknown. This is solved as previously:

$$3x + 84 - 8x = 44$$
$$3x - 8x = 44 - 84$$
$$-5x = -40$$

or
$$5x = 40$$

and
$$x = 8.$$

Substituting this value of x in equation (3) you can find y.

Thus
$$y = 21 - (2 \times 8)$$

∴
$$y = 5.$$

∴ the solution is $x = 8$, $y = 5$.

You should check these values by substitution in both of the given equations.

■ *Example 2:* In the following example a second method is shown which can frequently be employed to advantage.

Solve the equations:

$$x + y = 15 \qquad (1)$$
$$3x - y = 21 \qquad (2)$$

You can see that if the left sides of the two equations were added, the term in y would be eliminated, since you would get $(+y) + (-y) = 0$. Thus only x would remain.

It is clear also that the sum of the two left sides of the equations must equal the sum of the two right sides, i.e.,

$$(x + y) + (3x - y) = 15 + 21,$$

whence
$$x + 3x = 36$$
$$4x = 36.$$

∴
$$x = 9.$$

Substituting this value of x in equation (2) or, if easier, in equation (1)

$$(3 \times 9) - y = 21.$$

∴
$$27 - y = 21$$
$$-y = -6.$$

∴
$$y = 6.$$

∴ the solution is $x = 9$, $y = 6$.

■ *Example 3:* In the following example both of the above methods are employed.

Solve the equations:

$$2x + 3y = 42 \qquad (1)$$
$$5x - y = 20 \qquad (2)$$

1st method.   Substitution.

From (2) $\qquad\qquad -y = 20 - 5x.$
∴ $\qquad\qquad\qquad y = 5x - 20 \qquad (3)$

Substituting in (1):

$$2x + 3(5x - 20) = 42$$
$$2x + 15x - 60 = 42.$$
∴ $\qquad\qquad\qquad 17x = 102$
and $\qquad\qquad\qquad x = 6.$

Substituting for x in (3):

$$y = (5 \times 6) - 20$$
and $\qquad\qquad\qquad y = 10.$

∴ the solution is $x = 6, y = 10.$

2nd method.   Elimination.

In this example neither letter can be eliminated by addition of the left sides of the equation, as in example 2. But by multiplying both sides of equation (2) by 3, y can be eliminated.

You proceed as follows:

$$2x + 5y = 42 \qquad (1)$$
$$5x - y = 20 \qquad (2)$$

Multiplying by 3 throughout in (2)

$$15x - 3y = 60 \qquad (3)$$

Adding (1) and (3)

$$(2x + 3y) + (15x - 3y) = 42 + 60$$
∴ $\qquad\qquad\qquad 17x = 102$
and $\qquad\qquad\qquad x = 6.$

From this you can find y as before.

Note: You could have eliminated x from equations (1) and (2) as follows:

(a) Multiply throughout in (1) by 5.
(b) Multiply throughout in (2) by 2.

Then you get:
$$10x + 15y = 210$$
$$10x - 2y = 40$$

Subtracting

$$(10x + 15y) - (10x - 2y) = 210 - 40.$$
$$\therefore \quad 10x + 15y - 10x + 2y = 170$$
whence $\qquad\qquad\qquad 17y = 170$
and $\qquad\qquad\qquad\qquad y = 10.$

Of these two methods, that of substitution is the sounder and more general. In practice equations are seldom easily dealt with by the elimination method.

■ **Example 4:** Find values of R_1 and R_2 which will satisfy the equations:

$$0.5R_1 + 1.2R_2 = 1.486 \qquad\qquad (1)$$
$$4.5R_1 - 2R_2 = 4.67 \qquad\qquad (2)$$

From equation (1)

$$0.5R_1 = 1.486 - 1.2R_2.$$

$$\therefore \qquad R_1 = \frac{1.486 - 1.2R_2}{0.5}$$

$$= 2.972 - 2.4R_2.$$

Substituting in equation (2):

$$4.5(2.972 - 2.4R_2) - 2R_2 = 4.67$$
$$13.374 - 10.8R_2 - 2R_2 = 4.67$$
$$\therefore \qquad -12.8R_2 = -8.704$$

$$R_2 = \frac{8.704}{12.8}$$

$$\therefore \qquad R_2 = 0.68.$$

Substituting in equation (2):

$$4.5R_1 - (2 \times 0.68) = 4.67$$
$$4.5R_1 - 1.36 = 4.67$$
$$4.5R_1 = 6.03.$$

$$\therefore \qquad R_1 = \frac{6.03}{4.5}$$

$$= 1.34.$$

$\therefore$ the solution is $R_1 = 1.34$, $R_2 = 0.68$.

EXERCISE 21

Solve the following equations:

1 $y = 2x$
$3x + 2y = 21$

2 $y = 3x - 7$
$5x - 3y = 1$

3 $x = 5y - 3$
$3x - 8y = 12$

4 $x - y = 5$
$4x - y = 2x + 13$

5 $3x - 2y = 7$
$x + 2y = 5$

6 $2x - y = 3$
$x + 2y = 14$

7 $2x - y = 10$
$3x + 2y = 29$

8 $2(x - 4) = 3(y - 3)$
$y - 2x = -13$

9 $3x - 2(y + 3) = 2$

$2(x - 3) + 4 = 3y - 5$

10 $\dfrac{y}{2} - x = 2$

$6x - \dfrac{3y}{2} = 3$

11 $\dfrac{x}{2} - \dfrac{y}{5} = 1$

$y - \dfrac{x}{3} = 8$

12 $\dfrac{a}{2} + b = 1$

$3a - \dfrac{b}{3} = \dfrac{31}{2}$

13 $\dfrac{b}{3} - 2a = 5$

$3a + 4b = 6$

14 $4(1 - p) = 7q + 8p$

$6p + q + 8 = 0$

15 $7x + 3(y - 3) = 5(x + y)$

$7(x - 1) - 6y = 5(x - y)$

16 $\dfrac{x}{2} - \dfrac{y}{3} = \dfrac{1}{6}$

$\dfrac{y}{2} - \dfrac{x}{6} = 5$

17 $2(3a - b) = 5(a - 2)$
$3(a + 4b) = 2(b - 3)$

18 $0.1x + 0.2y = -0.2$
$1.5x - 0.4y = 10.6$

19 $1.25x - 0.75y = 1$
$0.25x + 1.25y = 17$

20 $2.5x + 3.7y = 13.365$
$8.2x - 1.5y = 7.02$

21 $\dfrac{x - 1}{3} + \dfrac{y + 2}{2} = 3$

$\dfrac{1 - x}{6} - \dfrac{y - 4}{2} = \dfrac{1}{2}$

22 $2P - 5Q = 2$

$3P + 10Q = 8.6$

23 $3(x - y) - \dfrac{1}{3}(x + y) = 30$ **24** $\dfrac{x}{3} + \dfrac{y}{2} = \dfrac{2x}{3} - \dfrac{y}{6} = 7$

$x + y + \dfrac{5}{3}(x - y) = 22$

64 PROBLEMS LEADING TO SIMULTANEOUS EQUATIONS

Many problems require for their solution the determination of two unknown quantities. The general method of solution is similar to that employed when there is one unknown, but with the important difference that when there are two unknowns to be determined, two equations must be formed from the data.

The following examples illustrate the methods employed.

65 WORKED EXAMPLES

■ *Example 1:* There are two numbers such that the sum of the first and three times the second is 53, while the difference between 4 times the first and twice the second is 2. Find the numbers.

Let x = one number.
Let y = the second.
Then from the first set of facts

$$x + 3y = 53 \qquad (1)$$

From the second

$$4x - 2y = 2 \qquad (2)$$

From (1) $x = 53 - 3y.$

Substituting in (2):

$$4(53 - 3y) - 2y = 2.$$
$$\therefore \qquad 212 - 12y - 2y = 2$$
Collecting $\qquad -14y = -210.$

$$\therefore \qquad y = \frac{-210}{-14}$$

and $\qquad y = 15.$

Substituting in (1) $\qquad x = 53 - 3y$
$$x = 53 - (3 \times 15).$$
$$\therefore \qquad x = 8.$$

∴ the numbers are 8 and 15.

■ *Example 2:* In the equation $y = mx + b$ it is known that the equation is satisfied by two pairs of values of x and y – namely when

$$x = 4, y = 6,$$
and when $\qquad x = 2.4, y = 4.5.$

What are the values of m and b?

This is an example of an important practical problem. It means that there is a law connecting x and y, the law involving m and b, which are constants. These constants must be determined before the law can be stated.

They are therefore the unknown numbers in this case. The equations connecting them are obtained by substituting the given pairs of values of x and y.

(1) When $x = 4$, $y = 6$. Therefore on substitution

$$6 = 4m + b \qquad\qquad (1)$$

(2) When $x = 2.4$, $y = 4.5$.

∴ $\qquad\qquad 4.5 = 2.4m + b \qquad\qquad (2)$

These are to be solved simultaneously for a and b. It is clearly a case for using the method of elimination.

Subtracting (2) from (1):

$$6 - 4.5 = (4m + b) - (2.4m + b)$$
or $\qquad 6 - 4.5 = 4m + b - 2.4m - b$
∴ $\qquad\qquad 1.5 = 1.6m.$

∴ $\qquad\qquad m = \dfrac{1.5}{1.6} \text{ or } \dfrac{15}{16}.$

Substituting in (1)

$$\left(4 \times \frac{15}{16}\right) + b = 6.$$

∴ $\qquad\qquad b = 6 - \dfrac{15}{4}$

and $\qquad\qquad b = \dfrac{9}{4}.$

∴ the solution is $\qquad m = \dfrac{15}{16}, b = \dfrac{9}{4}.$

Substituting these in

$$y = mx + b$$

you get $$y = \frac{15}{16}x + \frac{9}{4}$$

or clearing fractions $16y = 15x + 36.$

■ *Example 3:* A bookseller has a number of books the published price of which is £25. After selling a certain number at this price the bookseller sells the remainder at £20 each, and the total receipts are £1,100. If the numbers sold at the two prices were reversed, the bookseller would have received £1,150. How many books were there in all and how many were originally sold at £25?

Let x = the number originally sold at £25;
Let y = the number originally sold at £20.
The amounts received for these were $25x$ pounds and $20y$ pounds and their total value was £1,100.

$\therefore$ $$25x + 20y = 1100.$$

When the numbers are reversed the bookseller receives $20x$ and $25y$ pounds and their total value is now £1,150.

$\therefore$ $$20x + 25y = 1150$$

$\therefore$ the equations to be solved simultaneously are:

$$25x + 20y = 1100 \qquad (1)$$
$$20x + 25y = 1150 \qquad (2)$$

From (1) $$25x = 1100 - 20y$$

and $$x = \frac{1100 - 20y}{25}$$

Substituting in (2)

$$\frac{20(1100 - 20y)}{25} + 25y = 1150$$

$\therefore$ $$4(220 - 4y) + 25y = 1150$$
and $$880 - 16y + 25y = 1150$$
$\therefore$ $$9y = 270$$
$$y = 30$$

Since $$x = \frac{1100 - 20y}{25}$$

on substitution $$x = \frac{1100 - 600}{25} = \frac{500}{25}$$

∴ $$x = 20.$$

∴ the total number of books sold was $30 + 20 = 50$, and the number originally sold at £25 was 20.

EXERCISE 22

1 There are two numbers, x and y, such that the sum of $2x$ and y is 34, while the sum of x and $2y$ is 32. What are the numbers?

2 There are two numbers such that if to 3 times the first, twice the other is added, the sum is 72. Also if from 5 times the first number, 3 times the other is subtracted, the result is 44. What are the numbers?

3 One number is greater by 6 than twice another number, but 3 times the smaller number exceeds the greater by one. Find the numbers.

4 If from twice the greater of two numbers 17 is subtracted, the result is half the other number. If from half the greater number 1 is subtracted, the result is two thirds of the smaller number. What are the numbers?

5 In the equation $y = mx + b$, when $x = 3$, $y = 3$ and when $x = 5$, $y = 7$. Find the values of m and b and write down the equation. Then find y when $x = 6$.

6 Two quantities P and Q are connected by the formula:

$$P = \frac{m}{Q} + b$$

When $Q = 5$, $P = 14$ and when $Q = 2$, $P = 20$. Find m and b.

7 The force (E) applied to a machine and the resistance (R) to be overcome are connected by the law

$$E = a + bR.$$

It is found that when $E = 3.5$, $R = 5$ and when $E = 5.3$, $R = 8$. Find a and b. Then find E when $R = 10$.

8 It is known that $y = ax^2 + bx^3$; when $x = 2$, $y = 5.6$, and when $x = 3$, $y = 25$. Find the values of a and b.

9 The perimeter of a rectangular lawn is 32 m. It is reduced in size so that the length is four-fifths and the breadth is three-fourths of the original dimensions. The perimeter is then 25 m. What were the original length and breadth?

10 The bill for the telephone for a quarter can be expressed in the form

$$C = a + \frac{nb}{100}$$

where C is the total cost in pounds, a is a fixed charge, n the number of calls and b the price of each call in pence. When the number of calls was 104, the bill came to £58.30, and when the number was 67 the bill was £50.90. Find the fixed charge and the cost of each call.

11 The cost of 4 ties and 6 pairs of socks was £68.00, while that of 5 ties and 8 pairs of socks was £87.40. What were the prices of a tie and a pair of socks respectively?

12 The formula $S = ut + \frac{1}{2}at^2$ gives the distance S metres passed over by a moving body in t seconds.

In 4 seconds the body moves 88 metres.

In 6 seconds the body moves 168 metres.

Find the values of u and a and then find how far the body moves in 5 seconds.

9

Linear Inequalities

66 THE IDEA OF AN INEQUALITY

In many situations you use words like *is less than* or *at least*. Usually these will be statements involving numbers. For example:

'I have less than twenty pounds in my bank account'
'My ladder needs to be at least 4 m to reach the window'

It is not possible to deduce from these statements exactly how much you have in the bank or how long your ladder is. For this reason statements like this are sometimes called inequations, though it is more usual to describe them as **inequalities**. Special symbols are used to describe inequalities.

■ *Examples:*

4 *is less than* 7	$4 < 7$
3 m *is more than* 290 cm	$3 \text{ m} > 290 \text{ cm}$
My bank balance (B) *is less than* £20	$B < £20$
The ladder (L) *is at least* 4 metres	$L \geq 4 \text{ m}$

Note: the symbol $\geq$ is used to show that the length of the ladder is either *equal to* 4 m or is *more than* 4 m.
A similar symbol $\leq$ is used to show when one quantity is *less than* or *equal to* another quantity.

Each of the above statements can be rewritten using the symbol reversed.

$7 > 4$ 7 *is more than* 4

290 cm < 3 m   290 cm *is less than* 3 m
£20 > B   £20 *is more than* my bank balance
4 m ≤L   4 m *is less than or equal to L*

67 REPRESENTING INEQUALITIES

'My ladder is at least 4 m long, but must be shorter than 6 m if it is to fit into my garage.'

Suppose that the length of my ladder is L m. The above statement can be shown using the inequality symbols as:

$$L \geq 4 \text{ and } L < 6 \text{ or as the single statement } 4 \leq L < 6$$

One way to show the numbers which satisfy this inequality is to use a number line, as in Fig. 14.

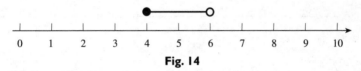

Fig. 14

On the number line the filled in ● is used to show that 4 can be included. However, since we were told that the ladder must be less than 6 m, an empty circle o is used at 6 to show that 6 m is not one of the possible lengths. The line joining the two circles indicates that the ladder can be any length between 4 m and 6 m.

■ *Example 1:* Represent the inequality on a 0 to 10 number line.

(a) $n < 3$ (b) $n \geq 6$ (c) $2 < n \leq 5$
(*a*) $n < 3$

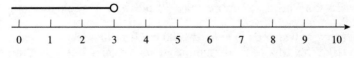

Fig. 15a

(*b*) $n \geq 6$

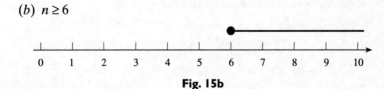

Fig. 15b

(*c*) $2 < n \leq 5$

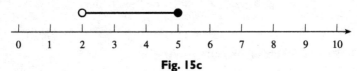

Fig. 15c

■ *Example 2:* What inequality is represented on the 0 to 10 number line?

(*a*)

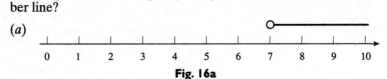

Fig. 16a

(*b*)

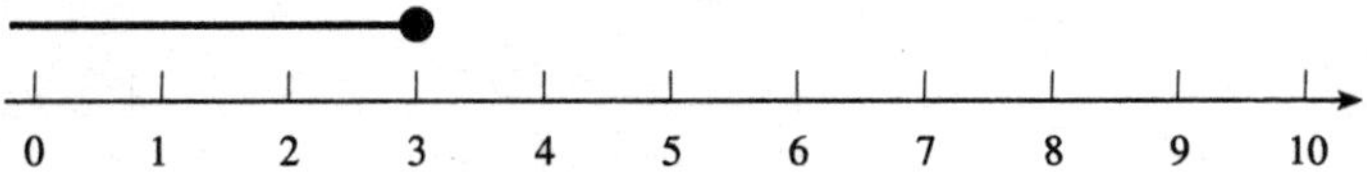

Fig. 16b

(*c*)

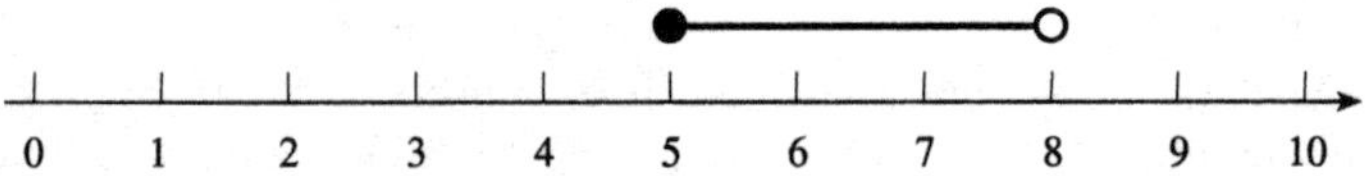

Fig. 16c

(*a*) represents $n > 7$, (*b*) represents $n \leq 3$, (*c*) represents $5 \leq n < 8$.

EXERCISE 23

1   Write as an inequality:

   (*a*) 'My height (*H*) must be less than 190 cm to be a pilot.'
   (*b*) 'She must sell more than 50 pies (*P*) a week to make a profit.'
   (*c*) 'I will spend (*S*) at least £200 but less than £300.'
   (*d*) 'The trip (*T*) takes more than 3 hours but less than 5 hours.'

2   Show the inequality on a 0 to 10 number line.

   (*a*) $n > 7$, where *n* is a whole number.
   (*b*) $n \leq 4$, where *n* is any number.
   (*c*) $3 \leq n \leq 8$, where *n* is a whole number.
   (*d*) $3 \leq n \leq 8$, where *n* is any number.

68 SOLVING INEQUALITIES

You can express many problems in the form of an inequality.

'Tickets for a concert cost £4, £5, £6, £7 or £8. I want to buy two tickets and I have £20 of which £7 is needed for the train fares.'

Assuming the two tickets are to cost the same this is a situation which can be represented by the inequality, $2C + 7 \leq 20$.

Clearly this is a problem which you can solve by checking the cost of each type of ticket. I could in fact buy two £4, or two £5 or two £6, but not two £7 tickets, since $2 \times 6 + 7 < 20$ whereas $2 \times 7 + 7 > 20$. However with more complex situations you need to have a more general technique. This technique is very similar to the one used for solving equations.

$$2C + 7 \leq 20$$

Subtracting 7 from each side gives

$$2C \leq 13$$

Dividing each side by 2 gives

$$C \leq 6.5$$

so C can be 4 or 5 or 6 but not 7.

You can think of an inequality as a balance where one side is heavier or lighter than the other. You can add or subtract the same number from each side without changing this situation. In the same way you could double or halve the numbers on each side, again without changing things.

However problems arise when you are dealing with negative numbers. It is helpful to look at a number line which shows both positive and negative numbers to illustrate this difficulty.

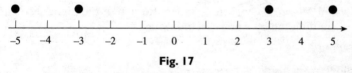

Fig. 17

The number line in Fig. 17 shows that the left-hand one of a pair of numbers on the line is smaller than the right-hand one of the pair. $3 < 5$, $-5 < 3$ and $-5 < -3$. It is important to realise that if $A > B$ then $-A < -B$.

In the first four examples below the inequality sign does not change.

$4 < 6$	adding 2 to each side gives	$4 + 2 < 6 + 2$
		i.e. $6 < 8$ ✓
$4 < 6$	subtracting 2 from each side gives	$4 - 2 < 6 - 2$
		i.e. $2 < 4$ ✓
$4 < 6$	doubling each side gives	$4 \times 2 < 6 \times 2$
		i.e. $8 < 12$ ✓
$4 < 6$	halving each side gives	$4 \div 2 < 6 \div 2$
		i.e. $2 < 3$ ✓

however if you were to multiply each side of $4 < 6$ by -2 you would get $4 \times (-2) = -8$ on the left side and $6 \times (-2) = -12$ on the right side. So in this case you need to change the inequality sign, since $-8 > -12$.

You can manipulate inequalities in the same way as equations, except that you need to remember that when you multiply or divide each side of an inequality by *a negative number* you must then change the sign of the inequality.

■ *Example 1:* Solve the inequality $3x - 5 < 2x + 8$

First add 5 to each side $\qquad 3x < 2x + 13$
Now subtract $2x$ from each side $\qquad x < 13$
So the solution is those values of x which are *less than* 13.

■ *Example 2:* Solve the inequality $3x + 5 > 5x - 9$

Method 1

Subtract 5 from each side $\qquad 3x > 5x - 14$
Subtract $5x$ from each side $\quad -2x > -14$
Divide each side by -2
  and change the sign $\qquad x < 7$

So the solution is those values of x which are *less than* 7.

Note: in *Method 1* you must change the sign of the inequality since you are dividing by a negative number.

Method 2

Add 9 to each side $\qquad 3x + 14 > 5x$
Subtract $3x$ from each side $\qquad 14 > 2x$
Divide each side by 2 $\qquad 7 > x$
Rewrite the inequality $\qquad x < 7$

So the solution is those values of x which are *less than* 7.

In *Method 2* the strategy was to collect the x terms on the side where they would be positive. Provided you are happy rewriting the inequality $7 > x$ as $x < 7$ this method is preferable and less likely to lead to mistakes.

■ *Example 3:* Solve the inequality $2(3x + 5) + 1 \geq 4x - 9$

First remove the brackets $6x + 10 + 1 \geq 4x - 9$
Now subtract 11 from each side $6x \geq 4x - 20$
Now subtract $4x$ from each side $2x \geq -20$
Finally divide each side by 2 $x \geq -10$

So the solution is those values of x which are *greater than* or *equal to* -10.

EXERCISE 24

1   Solve the inequality

  (a) $2x - 5 > 9$ (b) $3x + 7 < 16$

2   Solve the inequality

  (a) $5 > x - 3$ (b) $7 - x \geq 4$

3   Solve the inequality

  (a) $2x < x + 8$ (b) $5 + 2x \geq 3x$

4   Solve the inequality

  (a) $3x + 2 < x + 8$ (b) $2x + 7 \leq 5x + 1$

5   Solve the inequality

  (a) $7 - 2x > 3x - 8$ (b) $4 - 3x > 19 - 2x$

6   Solve the inequality $2(3x + 4) - 17 > 15$
7   Solve the inequality $3(x + 7) + 2 \leq 7(x - 2) + 5$
8   Solve the inequality $5(3x - 2) + 6 < 8(x - 7) + 3x - 12$

You can treat inequalities involving fractions in a similar way, but again it is very important to be careful and for you to be aware of when you might be multiplying or dividing by a negative number. The next example shows one of the possible hidden traps that sometimes occur.

■ *Example 4:* Solve the inequality $\dfrac{8}{x} < 2$

This inequality asks you to find the values of x which, when divided into 8, give a result which is less than 2. Clearly x can be any number which is greater than 4. For example $8 \div 5$ is 1.6 which is less than 2. However x could also be any negative number, since 8 divided by a negative number will also be a negative number and all negative numbers are less than 2.

If you were simply to multiply both sides of the inequality by x, this would give $8 < 2x$ or $x > 4$ as the solution. This however is only partly correct. Remember you can only multiply by x without changing the inequality sign if x is a positive number. If x were negative, which is the other part of the solution, then the inequality would become $8 > 2x$ which is certainly true.

69 SIMULTANEOUS INEQUALITIES

In Chapter 8 you were introduced to the idea of a linear equation which involved two variables, and also how to find the solution for a pair of such simultaneous linear equations. The corresponding work with inequalities involving two variables comes later in this book. However, here are some examples in which you might need to satisfy both of two simple linear inequalities at the same time.

■ *Example 1:* Find the values of x which satisfy both these inequalities

$$3x + 2 > 8 \text{ and } 5x - 3 < 27$$

Solving the first you get: $3x > 6$ or $x > 2$
Solving the second you get: $5x < 30$ or $x < 6$

So x must be a number which is *both greater than* 2 and *also less than* 6. So x can be any number *between* 2 and 6 but not including either 2 or 6.

You can also show this as the single statement $2 < x < 6$.

Using a number line the required solution is shown in Fig. 18.

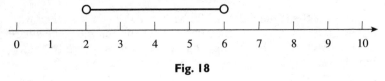

Fig. 18

■ *Example 2:* Find the positive values of x which satisfy both these inequalities.

$$3(x + 5) \geq 8x \text{ and } 5(x - 3) < 27 - x$$

Solving the first you get: $3x + 15 \geq 8x$ or $15 \geq 5x$ or $x \leq 3$
Solving the second you get: $5x - 15 < 27 - x$ or $6x < 42$ or $x < 7$

So x must be a positive number which is *less than or equal to* 3 and which is also *less than* 7. The only possible values of x are those which are *greater than or equal to* 0 and *less than or equal to* 3. So $0 \leq x \leq 3$.

Using a number line the required solution is shown in Fig. 19.

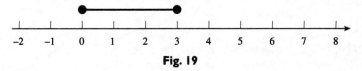

Fig. 19

EXERCISE 25

1 Find the positive values of x which satisfy both the inequalities:

$$2x < x + 8 \text{ and } 5 + 2x \geq 3x$$

2 Find the positive values of x which satisfy both the inequalities:

$$3x - 2 < x + 8 \text{ and } 2x + 7 \leq 5x + 1$$

3 Find the positive values of x which satisfy both the inequalities:

$$3(x + 5) < 8x \text{ and } 5(x - 3) < 27 - x$$

With more complex situations it may be necessary to use a number line to show which numbers satisfy both inequalities.

■ *Example 3:* Find the values of x which satisfy both inequalities.

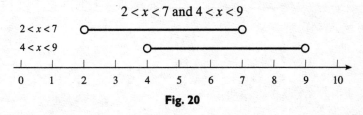

Fig. 20

The numbers which satisfy both inequalities at the same time are the numbers between 4 and 7, i.e. $4 < x < 7$.

■ *Example 4:* Find the values of x which satisfy both inequalities.

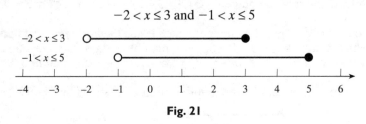

$$-2 < x \leq 3 \text{ and } -1 < x \leq 5$$

Fig. 21

The numbers which satisfy both inequalities at the same time are the numbers between -1 and 3, including 3 but excluding -1, i.e., $-1 < x \leq 3$.

EXERCISE 26

1 Find the values of x which satisfy both inequalities:

   (*a*) $4 < x < 8, 3 < x < 7$ (*b*) $-2 < x < 1, -3 < x < 7$

2 Solve the simultaneous inequalities: $2(3x - 4) < x + 7$ and $-3 \leq x < 5$.

10

Graphical Representation of Quantities

70 THE OBJECT OF GRAPHICAL WORK

The graphical or pictorial representation of statistics and the comparison of magnitude by means of various graphical devices are familiar features of modern life. Rows and columns of figures or groups of large or very small numbers are not always readily grasped.

Accordingly, the use of lines or columns or other figures drawn to scale which appeal to the brain through the eye, is found to be an effective method of enabling many people to understand not only the quantities themselves but also to make deductions from them.

The graphical method, developed on scientific and mathematical lines, is also largely employed in mathematics and science to illustrate certain important underlying principles. The following examples, arranged progressively, will help you to understand the various forms of graphical representation.

71 THE BAR GRAPH

The first example, Fig. 22, is a reproduction of an actual graphical advertisement, issued by the Cement Makers' Federation to show the variations in the deliveries of cement over a period of 11 years. It is an example of a **bar graph**.

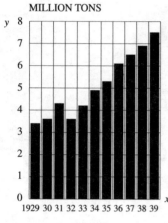

MILLION TONS

Fig. 22

Along the horizontal base line, called the **x-axis**, equal spaces are marked off which indicate, in succession, the eleven years from 1929 to 1939 inclusive.

In the spaces so formed are constructed a series of columns of the same width and whose height, on the scale chosen, represents the number of tons of cement manufactured in each year. In order that these heights can be easily read a scale is marked along the vertical line, called the **y-axis**, perpendicular to the **x-axis**, in which each unit represents a million tons.

These columns convey very effectively, in a general way, the rise and fall of the production of cement during the eleven years. It also enables you to tell at a glance the approximate amount of cement produced in any one of the years.

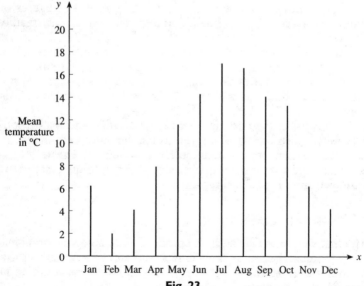

Fig. 23

72 A STRAIGHT-LINE GRAPH

The next method illustrated is an alternative to the bar graph. The following table shows the mean temperature for each month throughout the year.

(1969) Month	Jan.	Feb.	Mar.	Apr.	May	June	July	Aug.	Sept.	Oct.	Nov.	Dec.
Mean temp. in °C	6.2	2.0	4.1	7.9	11.6	14.3	17.0	16.6	14.1	13.3	6.2	4.2

These are exhibited graphically in Fig. 23, and the method is as follows. Draw the *x*-axis, and at suitable equal intervals mark points corresponding to the twelve months.

Draw the *y*-axis at right angles to the *x*-axis, and with a suitable scale mark points corresponding to temperatures from 0° to 20°.

To save time and secure accuracy, you may find it convenient to use specially ruled paper, called 'graph paper' or a computer.

At each of the 12 points on the *x*-axis draw lines perpendicular to the *y*-axis to represent the corresponding temperature as indicated on the scale on the *y*-axis.

These straight lines take the place of the columns in Fig. 22.

This diagram shows at a glance not only the mean temperature for each month, but gives a clear picture of the rise and fall of average temperature throughout the year, the highest and the lowest.

73 A GRAPH

You can see that when graph paper is used it is not necessary to draw the lines perpendicular to the *x*-axis (as was done in Fig. 23). A point will mark the top of such a line, if it were drawn.

You now join these points by straight lines as shown in Fig. 24. The succession of lines thus formed is called a **graph**.

This is much more useful than the series of perpendicular lines in Fig. 23, and lends itself to important developments, as you will see in later examples. Not only does it show more vividly the rise or fall each month, the highest and lowest temperatures, etc., but other features are apparent. For example, you get an idea of the average rate at which temperature is rising or falling during each month. This is indicated by the slope of the line corresponding to the month. Between November and December the slope of the line is very slight, the drop in temperature being small. Between March and April the steep slope shows a rapid increase in temperature. The month of the sharpest rise is April, when the slope is greatest. The slope is reversed when the temperature falls, and the month when the fall in temperature is greatest is evidently October. January seems to have been an exceptionally warm month in 1969; the oddness of the temperature is reflected in the oddness of the graph, which kinks at this point.

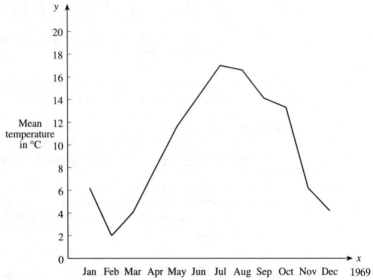

Fig. 24

74 EXAMPLES OF GRAPHS AND THEIR USES

Graphs are employed in almost every branch of knowledge, and a full treatment of them is impossible in a book of this size. The following graphs may serve to illustrate their nature and their uses. The first example is from Insurance.

■ *Example 1:* The annual premiums charged by an Insurance Company for a policy of £100, at various ages, are as follows:

Age:	25	30	35	40	45	50
Premium in £s	2.33	2.59	2.91	3.31	3.81	4.53

The amounts of the premiums have been taken to the nearest penny. The method of 'plotting' these values, as it is called, is as follows.

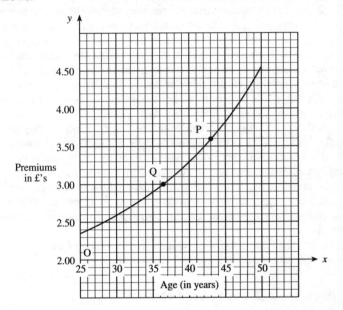

Fig. 25

The x- and y-axes are drawn at right angles (Fig. 25). Ages are marked off on the x-axis, one small division representing a year. These are named as shown. Premiums are marked on the y-axis, one small division representing 10p.

The point where the x- and y-axes meet is called the **origin**, and is usually labelled O. Sometimes the x-axis is called Ox and the y-axis is called Oy.

As there is no premium less than £2, begin marking the premiums by placing the mark for £2 at O; similarly on the x-axis begin the ages with twenty-five years at O. By this means space on the paper is used to best advantage.

In succession find the points which show the premium corresponding to the ages given. These are marked with dots. You should take squared paper and get these points for yourself. The process of obtaining the points is called 'plotting the points'.

When this is done you can see that the points appear to lie on a regular, smooth curve, and this has been drawn on Fig. 25. An important point now arises.

If it is correct to assume that points corresponding to the premiums for every fifth year lie on a regular smooth curve, then the values of these would seem to have been calculated according to a definite law or formula. If that is the case, then the premiums for the intervening years, which are not given in the table, would be calculated according to the same formula. It is therefore a reasonable deduction that the points corresponding to these, if plotted, would lie on the curve already obtained.

Accordingly, if you note the point of intersection of the curve with the line parallel to the y-axis corresponding to any intervening year, the position of this point with reference to the scale on the y-axis indicates the amount of the premium for the year selected.

Thus, considering the point P, corresponding to the forty-third year, the premium as shown on the y-axis is £3.60. Conversely, when the premium is £3 at Q, the corresponding age is just over thirty-six.

This process of obtaining values from the graph which lie between those plotted is called **interpolation**.

75 AN EXAMPLE FROM ELECTRICITY

The table below shows the resistance in ohms for given lengths of wire of the same material and cross-section.

Length in metres	100	120	170	220
Resistance in ohms	2.5	3	4.25	5.5

Draw the graph which shows the relation between resistances and length and find the resistance for a length of 200 m.

The x- and y-axes are drawn as before (Fig. 26).

Lengths are marked along the x-axis, beginning with 100 m at the origin O, and each small division representing 5 m.

Resistances are marked along the y-axis, five divisions representing 1 ohm and

∴ each division representing 0.2 ohm.

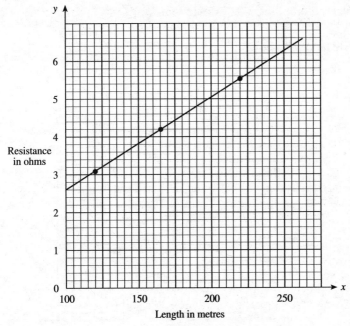

Fig. 26

Using corresponding values from the table, plot the points as before. You can see that these points lie on a straight line.

As in the preceding example, you might assume that a definite law connects resistance and length.

Using the method of interpolation you can see that when the length is 200 m the resistance is 5 ohms.

The graph being a straight line, you can produce it as in the figure. This enables you to find the resistances corresponding to lengths beyond those given.

For example, when the length is 250 m the resistance is a little greater than 6.2 ohms.

This is called **extrapolation**.

76 AN EXAMPLE FROM MECHANICS

The following table shows the distances passed over in the times indicated by a body starting from rest. Draw a graph to show the relation between time and distance.

Algebra

Time in seconds	0	1	2	3	4	5
Distance in metres	0	2	8	18	32	50

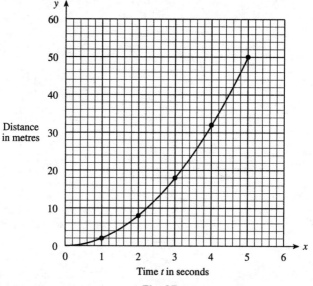

Fig. 27

Drawing the x- and y-axes, units are selected in accordance with the range of numbers to be represented (see Fig. 27).

The points plotted are indicated by small dots. They appear to lie on a smooth curve, and this is drawn to include all the points given in the table.

As in previous examples, the smoothness of the curve suggests that time and distance are connected by a definite formula. This is $y = 2x^2$. This will be familiar to those students who have studied mechanics.

The curve is called a parabola. Reasoning as in the previous example you can use the curve for interpolation. The following examples illustrate this.

■ *Example 1:* Find the distance passed over in 3.6 s.

From the curve this appears to be 26 m. The actual distance found from the formula $y = 2x^2$ is 25.92 m.

■ *Example 2:* How long will the body take to travel 42 m?

Finding the point which marks 42 m on the *y*-axis, and then look-ing for the corresponding point on the curve, this is seen, from the scale on the *x*-axis, to represent 4.6 s.

The Mechanics formula gives 4.58 s.

EXERCISE 27

1   The following table gives the values of exported manufactured goods of a certain type in years as specified.

Year:	1990	1991	1992	1993	1994	1995
Value in million £s	6.0	5.2	4.9	5.1	4.8	5.6

Show the variations by means of a graph.

2   The temperatures between 1000 h and 2000 h taken every two hours were as follows:

Time	1000	1200	1400	1600	1800	2000
Temp °C	8	9.1	11.9	13.7	12.1	9.0

Show these in a graph and find the probable temperature at 1300 h.

3   The expectation of life in years for males and females in this country at various ages is shown in the following table:

Age:	20	30	40	50	60	70
Expectation (Males)	47.2	38.5	29.8	21.8	14.7	8.8
Expectation (Females)	50.3	41.6	32.8	24.4	16.7	10.2

From your graph estimate the expectation of life for males and females at (1) 35 years, (2) 55 years.

4   Information about the average weight of children of different ages is given in the following table. Draw a graph to illustrate the information.

Age (years):	11	12	13	14	15
Weight (kg)	40	42.5	46	50.5	57

(1) Use your graph to find an estimate for the average
weight of a child of age 12½ years.
(2) Use your graph to find an estimate for the average
weight of a 16 year old.
(3) If a child of age 13½ years has a weight of 46.5 kg, by how
much is this child below average?

5 The populations of Salford CB and Poole MB at certain times
is shown in the table below. Draw, using the same axes and
scales, graphs showing the change in the two populations.

(1) From your graphs, when are the populations likely to be
equal?
(2) What will this equal population be?

Draw the graph showing the difference in the two populations.

(3) When does this show the difference as being zero?

[*N.B.* In order to use the same axes and scales as before, draw
the graph of 'difference + 100 000': the 'zero-line' will then be
100 000.]

Population in thousands

Year	1951	1961	1966	1968	1969
Salford	177	154	144	140	138
Poole	83	92	97	99	102

6 A train starts from rest, and its speed at intervals during the
first minute is given in the following table:

Time (s)	0	5	10	20	30	40	50	60
Speed (m/s)	0	8.5	14.6	23	29.2	33.6	37	39

Draw a graph showing the relation between time and speed.
Does it appear that a definite law connects these?
    What is your estimate for the speed after 23 s?
7 The following table gives the annual premium £P payable during life by a person aged A years next birthday for a whole life
assurance of £10,000. Draw the graph.

A	20	25	30	35	40	45	50	55
P	116	131	133	175	210	255	315	400

A mistake was made in working out these figures: find from your graph which entry is wrong and estimate the correct entry. Estimate the annual premium for people aged 24, 42, 53 next birthday.

8 The following table shows the sales of gas for public lighting in the North West in the years 1963/4–1968/9. What is the likely sale in 1969/70?

Gas sales in kilotherms

Year	63/4	64/5	65/6	66/7	67/8	68/9
Gas	1933	1466	1017	723	552	434

9 If $y = \frac{1}{2}(x - 3)(x + 1)$, work out values of y for successive whole number values of x from $x = -3$ to $x = 5$. Draw a graph of $y = \frac{1}{2}(x - 3)(x + 1)$. From your graph, read off the two values of x that make y equal to 1.

11

The Law of a Straight Line; Coordinates

77 THE STRAIGHT-LINE GRAPH

Among the graphs considered in the previous chapter were some which were straight lines. This regular arrangement of the plotted points suggested that a definite law governed the relations between the quantities, corresponding values of which were the basis of the plotting. This is an important principle which calls for further investigation.

The following example will serve as a starting-point.

The proprietor of a restaurant calculated the following figures showing the connection between the net profits and the number of customers. Exhibit the connection by a graph.

Number of customers	24	27	30	35	38
Net profit in £s	5	20	35	60	75

Note: 'Net profit' means total receipts from customers, less expenses.

Let x = the number of customers.

Let £y = the net profit.

Choosing the two axes as usual, values of x will be plotted on the x-axis, values of y on the y-axis.

When plotted as in Fig. 28, the points are found to lie on a straight line.

From this point in the book, the graphs will be drawn as in Fig. 28, without squared paper, to make them clearer to follow. You should still use squared paper in your own work.

Interpolation and extrapolation can be used as previously, but two important questions arise:

(1)   What is the number of customers when there is no profit?

    Zero on the profit (y) scale is shown where the graph cuts the x-axis. Producing the graph it cuts at the point (A), where $x = 23$, i.e.,

when $x = 23, y = 0.$

(2)   What happens when the number of customers is less than 23?

    If there is no profit when the number of customers is 23, there will be a loss when the number is below 23.

Extrapolate the straight line below the x-axis, as in Fig. 28. The amount of the loss, as with the profit, will be shown on the y-axis, which must also be extrapolated.

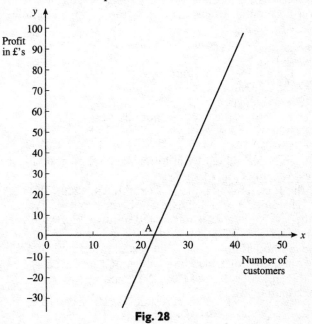

Fig. 28

The question of indicating loss as contrasted with profit now arises as it did in Chapter 4. You must proceed in the same way. The loss will be marked with negative numbers on the scale, as shown. You can then see that on the *y*-axis there is thus constructed part of a complete number scale (see Section 38).

From this scale, it appears that when *x* (the number of customers) is 18, *y* (the loss) is £25.

Problems will also arise when it will be necessary to use negative values of *x*, on the *x*-axis produced. Consequently for a complete graph you need to complete number scales on both axes, with zero common to both.

In the particular example above, a negative number of customers would have no intelligible meaning.

78 THE LAW REPRESENTED BY A STRAIGHT-LINE GRAPH

The next step is to discover the nature of the law which a straight-line graph represents and how it can be formulated. Using the above problem as an example, it was stated that the net profit was equal to the total receipts less the expenses. But the total receipts equal (the number of customers) × (the average amount paid by each).

Let £a = the average amount thus paid.
Then since
x = the number of customers,
ax = the total amount paid in pounds.
Let £b = the expenses.
Then net profits are ($ax - b$) pounds, i.e.

$$y = ax - b.$$

This gives the value of *y* in terms of *a*, *x*, and *b*, and is the form of the law connecting them.

Of the four letters, *a* and *b* remain unchanged, while the number of customers (*x*) and consequently the profit (£*y*) vary.

∴ the law is not completely stated until we know the values of *a* and *b*.

The method of doing this is suggested by example 2, Section 65.

Two pairs of corresponding values of *x* and *y* can be obtained from the table of values or from the graph.

For example, when

$$x = 32, y = 45$$

when $\qquad\qquad x = 25, y = 10.$

Substituting these in the equation $y = ax - b$ you get

$$45 = 32a - b \qquad\qquad (1)$$
$$10 = 25a - b \qquad\qquad (2)$$

Subtracting equation (2) from equation (1)

$$35 = 7a$$

and $\qquad\qquad a = 5.$

Substituting in equation (1) you get

$$b = 115.$$

$\therefore$ the equation is $\qquad\qquad y = 5x - 115.$

This is the law of the straight line in Fig. 28.

If it is correct it must be satisfied by any corresponding pair of values of x and y, and you should test this out.

In particular if $y = 0$, i.e. there is no profit

$$5x - 115 = 0$$

whence $\qquad\qquad 5x = 115$

and $\qquad\qquad x = 23.$

This agrees with the result found above.

Thus it can be demonstrated that the equation

$$y = 5x - 115$$

is satisfied by the coordinates of any point on the line and so is called the **equation of the straight line**.

79 GRAPH OF AN EQUATION OF THE FIRST DEGREE

The equation which is represented by the straight line in the above problem (namely, $y = 5x - 115$) is of the first degree, that is, it contains no higher powers than the first of x and y.

Two questions now suggest themselves.

(1) Can every equation of the first degree in x and y be represented graphically by a straight line?
(2) Conversely, can every straight line be represented algebraically by an equation of the first degree?

The answers to these questions will be apparent later, but for the present only the graphs of some typical equations of the first degree will be drawn. From these graphs you can deduce the answer to the first question.

Examples are given to illustrate the methods to be employed. In all of these, since both positive and negative values of x and y will be involved, complete number scales will be used on both axes.

Before calculating corresponding values of x and y, you should revise Section 62 if you are not familiar with the procedure.

80 WORKED EXAMPLES

■ *Example 1:* Draw the graph of the equation:

$$2y - 4x = 3.$$

This is not in the form used above, but you can transform it into the same form.

$$2y - 4x = 3$$
$$2y = 4x + 3.$$
$$\therefore \qquad\qquad y = 2x + 1.5.$$

Giving suitable values to x, you can calculate the corresponding values of y, and get the following table:

x	-2	-1	0	1	1.8	2
y	-2.5	-0.5	1.5	3.5	5.1	5.5

A straight line is fixed by two points, but in drawing it from its equation you should always take three points as a check on accuracy. In this example, a number of points are taken, so as to make it clear that all such points lie on the straight line. The graph appears as shown in Fig. 29.

You could check this by finding for any point on the line the corresponding values of x and y. These values should satisfy the equation $y = 2x + 1.5$.

Intercepts.
Notice that when

$$\text{(1) } x = 0, y = 1.5,$$
when $\qquad\qquad \text{(2) } y = 0, x = -0.75.$

(1) 1.5 is called the intercept on the y-axis ($x = 0$).
(2) -0.75 is called the intercept on the x-axis ($y = 0$).

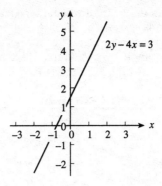

Fig. 29

■ *Example 2:* Draw the graph of the equation:

$$2x + y = 1$$

Transforming the equation you get:

$$y = -2x + 1 \tag{A}$$

You then construct a table of corresponding values.

x	-2	-1	0	1	2	3
y	5	3	1	-1	-3	-5

With these values the straight line shown in Fig. 30 is drawn.

For intercepts on the axes when $x = 0$, $y = 1$ (intercept on y-axis)

when $y = 0$, $x = 0.5$ (intercept on x-axis).

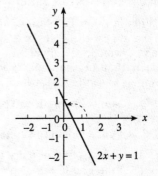

Fig. 30

Note from this example that when the coefficient of x in the equation arranged as in (A) is negative, the angle which the straight line makes with the x-axis in an anti-clockwise direction is greater than a right angle. When the coefficient is positive, as in Fig. 29, this angle is less than a right angle.

EXERCISE 28

Draw the graphs which are represented by the following equations. In each case find the intercepts on (1) the y-axis, (2) the x-axis.

1 $y = x + 2$ **2** $y = 1.5x - 1$

3 $4y = 6x - 5$ **4** $3x + 2y = 6$

5 $\dfrac{x}{3} + \dfrac{y}{2} = 2$ **6** $2(x - 3) = 4(y - 1)$

7 The equation $y = ax + b$ is satisfied by the following pairs of values of x and y:

    (1) $x = 1, y = 5$
    (2) $x = 2, y = 7$

Find the values of a and b, and substitute in the equation. Draw the graph of this equation and find its intercept on the y-axis.

8 A straight line makes an intercept of 2 on the y-axis and of 4 on the x-axis. Find its equation.

81 POSITION IN A PLANE; COORDINATES

When a point is 'plotted', as in graphs previously considered, its position on the graph was fixed by the corresponding values of x and y which were given in the table. For example, let P be a point such that in the table $x = 2$, $y = 3$. When plotted, the point appears as in Fig. 31, where:

PQ is 2 units in length and parallel to the x-axis,
and
PR is 3 units in length and parallel to the y-axis.

The intersection of these two straight lines fixes the position not only in the graph, but also relative to the x- and y-axes. The position of any other point can be similarly determined when its distances from the two axes are known.

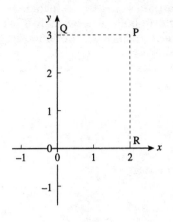

Fig. 31

You could use this method to find where, in a field, you had hidden something. By remembering the distances OR and OQ along two boundaries of the field, you can 'plot the point P' in the field. A little reflection will convince you that the boundaries need not be at right angles to one another.

The distances PQ and PR (Fig. 31) which thus fix the position of a point are called the **coordinates** of P with respect to the two axes.

PQ, parallel to the x-axis, is called the x-coordinate (or the abscissa).

PR, parallel to the y-axis, is called the y-coordinate (or the ordinate).

The notation employed to denote coordinates is (2, 3) or in general (x, y). The x-coordinate is always placed first inside the bracket.

Thus (5, 2) would represent the point in which $x = 5$, $y = 2$. In this way the positions of points relative to two axes can be

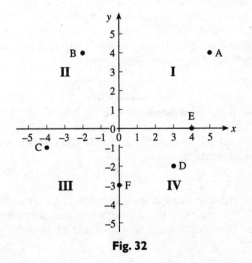

Fig. 32

described concisely. If, however, all points are to be included, complete number scales must be used on both axes, as shown in Fig. 32. In this way four divisions, called quadrants, are found in the plane; numbered I, II, III, IV as shown.

The signs of the coordinates are regulated by the positive and negative parts of the number scales.

The rule of signs is as follows:

If (x, y) are the coordinates of a point:

> x is +ve when measured to the right of O.
> x is −ve when measured to the left of O.
> y is +ve when measured up from O.
> y is −ve when measured down from O.

As examples the coordinates of A, B, C, D in Fig. 32 are:

> A is $(5, 4)$, B is $(-2, 4)$, C is $(-4, -1)$, D is $(3, -2)$.

Points on the axes.

If a point is on the x-axis the y-coordinate is zero. Thus E is $(4, 0)$.

If a point is on the y-axis the x co-ordinate is zero. Thus F is $(0, -3)$.

As the origin O lies on both axes, its coordinates are $(0, 0)$.

Thus the position of every point on a plane relative to two axes can be determined by coordinates.

Latitude and longitude are a practical example of the use of coordinates. They describe the position of a place with reference to the equator and the meridian through Greenwich as axes.

The introduction of coordinates was due to Descartes, who published his book on Analytical Geometry in 1638.

EXERCISE 29

1 Write down the coordinates of the points in Fig. 33 marked A, B, C . . . G.

2 Join OA and FC in Fig. 33, and find the coordinates of their point of intersection.

3 Plot the points $(3, 1)$, $(1, 3)$, $(0, 3)$, $(0, -3)$, $(4, 2.6)$, $(-2, 1)$, $(-4, -2)$, $(3, 0)$, $(-2, 0)$.

4 Draw the straight lines joining the points $(3, 1)$ and $(1, 3)$ and also $(-2, 1)$ and $(4, 2.6)$ as plotted in the previous question. What are the coordinates of the point of intersection of the two lines?

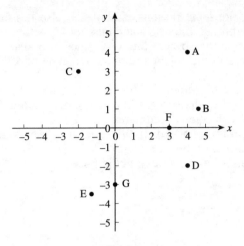

Fig. 33

5   Plot the points $(-3, 2)$, $(0, 2)$, $(2, 2)$, $(4, 2)$. What do you notice about these points?

6   Plot the points $(3, 3)$, $(1, 1)$, $(-1, -1)$, $(-2, -2)$. What do you notice about these points?

7   Draw a straight line through $(3, 0)$ parallel to the *y*-axis. What do you notice about the coordinates of points on this line?

82 A STRAIGHT LINE AS A LOCUS

Let A (Fig. 34) be a point such that its coordinates are equal.
    Let (x_1, y_1) be its coordinates.

Then $y_1 = x_1$.

Join OA and draw AP perpendicular to the *x*-axis.

Then $AP = OP$.

∴ by geometry $\angle OAP = \angle AOP$

since ∟APO is a right angle.
    ∴ each of the angles OAP, AOP is 45°.
    ∴ A lies on a straight line passing through the origin and making 45° with the *x*-axis.
    Let B be any other point with equal coordinates (x_2, y_2) so that $BQ = OQ$, i.e., $y_2 = x_2$.

Then, for the same reasons as above, B also lies on a straight line passing through the origin and making 45° with the *x*-axis.

This must be the straight line OA, since only one straight line can pass through the origin and make an angle of 45° with the *x*-axis.

Similarly all other points with equal coordinates lie on the same straight line – i.e., the straight line OA, extrapolated, is the locus of all points with equal coordinates.

These coordinates all satisfy the equation $y = x$, which is therefore the equation of the line.

This equation can be written in the form

$$\frac{y}{x} = 1.$$

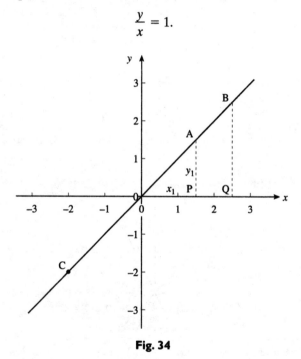

Fig. 34

∴ for any point such as A,

$$\frac{AP}{OP} = 1.$$

This ratio is constant for all points on the straight line, and is called the gradient of the line.

A similar result holds for every straight line, consequently a straight line is a graph which has a constant gradient.

To include all points with equal coordinates the straight line must be extrapolated into the opposite (3rd) quadrant.

Then for any point on this line the coordinates are equal, but they are both negative. Thus for the point C (Fig. 34) these coordinates are $(-2, -2)$ and the gradient is $\dfrac{-2}{-2}$, i.e., unity.

You may know, from geometry or trigonometry that the ratio of $\dfrac{AP}{OP}$ is constant for any point A. See also Section 169.

83 EQUATION OF ANY STRAIGHT LINE PASSING THROUGH THE ORIGIN

The conclusions reached above apply equally to all straight lines through the origin. The lines differ only in the gradient – i.e., in

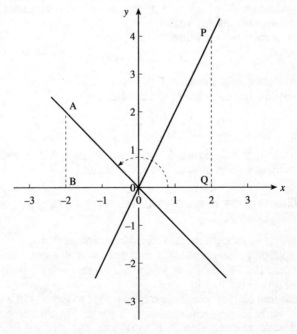

Fig. 35

the value of $\dfrac{y}{x}$.

For example, if the gradient is 2, then $\dfrac{y}{x} = 2$ and $y = 2x$.

This is shown in Fig. 35. For any point P on the line OP

the ratio $\dfrac{PQ}{OQ} = \dfrac{4}{2} = 2.$

Generally, if the gradient is denoted by m then

$$\frac{y}{x} = m.$$

$\therefore$ $y = mx.$

This is the general form of the equation of any straight line through the origin, where m denotes the gradient.

Negative gradient. If m is negative, the line will pass through the 2nd and 4th quadrants.

Thus if $m = -1$ the equation is

$$y = -x.$$

The straight line is shown in Fig. 35.

Considering any point A, the gradient is

$$\frac{AB}{OB} = \frac{+2}{-2} = -1.$$

Note that the angle made with the x-axis is 135°, angles being always measured in an anti-clockwise direction.

84 GRAPHS OF STRAIGHT LINES NOT PASSING THROUGH THE ORIGIN

In Fig. 36 the straight line AOB is the graph of $y = x$.

If you plot the graph of $y = x + 2$, it is evident that for any value of x the value of y in $y = x + 2$ is greater by 2 than the corresponding value of y in $y = x$.

$\therefore$ the line for $y = x + 2$ must be parallel to $y = x$, but each point is two units higher in the y scale. Thus in Fig. 36 the point A is raised to A′, the origin to D, B to B′, etc.

A′DB′ therefore represents the graph of $y = x + 2$.

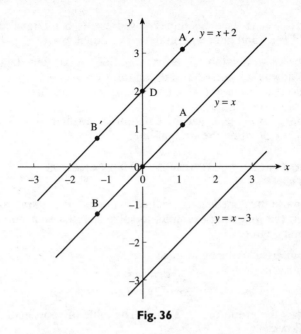

Fig. 36

The straight line $y = x + 2$ is the locus of all points whose coordinates are such that the y-coordinate = the x-coordinate + 2. It has the same gradient as $y = x$, but its intercept on the axis of y is +2.

Similarly the line $y = x - 3$ is parallel to the line $y = x$, with each point on it lowered by 3 units in the y scale.

You can generalise from this and state that $y = x + b$ will always represent a straight line parallel to $y = x$ and with an intercept of b units on the y-axis for any value of b.

The same conclusions hold for lines with different gradients. For example, the equation $y = 2x + b$ will always represent a straight line parallel to $y = 2x$, i.e., having the same gradient, and with an intercept of b units on the y-axis.

Generalising, let m = the gradient of a straight line.

Then $y = mx + b$ always represents a straight line parallel to $y = mx$, i.e., with the same gradient, and with an intercept of b units on the y-axis.

■ *Examples:*

(1) $y = 4x - 7$ is a straight line whose gradient is 4, and whose intercept on the y-axis is -7.

(2) $y = 5x - 115$ (see Section 78) represents a straight line of gradient 5 and intercept on the y-axis equal to -115.

As shown in Section 80, every equation of the first degree in two unknowns can be reduced to the form $y = mx + b$.

It follows that the graph of every equation of the first degree in two unknowns is a straight line.

Further, it is evident that the equation is satisfied by the coordinates of any point on the straight line.

85 GRAPHICAL SOLUTION OF SIMULTANEOUS EQUATIONS

You can use the conclusions reached in the previous paragraph, as shown in the following example, to solve simultaneous equations of the first degree.

■ *Example:* Solve the equations:

$$x + 2y = 5 \qquad (1)$$
$$3x - 2y = 7 \qquad (2)$$

(1) Draw the graph of $x + 2y = 5$. The table of coordinates is as follows:

x	0	-1	2	5
y	2.5	3	1.5	0

Note: When $x = 0$, y is the intercept on the x-axis; when $y = 0$, x is the intercept on the x-axis. It is useful to obtain these two points.

The graph is the straight line marked A in Fig. 37.

(2) Draw the graph of $3x - 2y = 7$. The table of values is:

x	0	-1	$\frac{7}{3}$	4
y	-3.5	-2	0	2.5

The graph is the straight line marked B.

Applying the conclusions, reached in Section 84:

Line A contains all those points whose coordinates satisfy the equation $x + 2y = 5$.

Line B contains all those points whose coordinates satisfy the equation $3x - 2y = 7$.

There is one point, and one point only, whose coordinates satisfy both equations.

That point is P, the intersection of the two graphs.

The coordinates of P, by inspection, are (3, 1).

∴ the solution of the equation is

$$x = 3, y = 1.$$

You should compare the above conclusions with the algebraical treatment in Section 62.

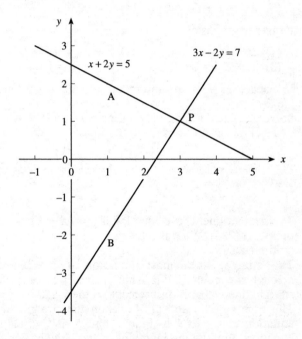

Fig. 37

EXERCISE 30

1   With the same axes draw the graphs of:

(1) $y = x$ (2) $y = 2x$ (3) $y = \frac{1}{2}x$
(4) $y = -x$ (5) $y = -2x$

2 With the same axes draw the graphs of:

(1) $y = x$ (2) $y = x + 1$ (3) $y = x + 3$
(4) $y = x - 1$

3 With the same axes draw the graphs of:

(1) $y = \frac{1}{2}x$ (2) $y = \frac{1}{2}x + 2$ (3) $y = \frac{1}{2}x - 1$

4 With the same axes draw the graphs of:

(1) $y = x + 2$ (2) $y = 2x + 2$
(3) $y = -x + 2$ (4) $y = \frac{1}{2}x + 2$

5 Draw the graphs of:

(1) $2x + y = 3$ (2) $x - 2y = 4$
(3) $5x + 2y = 10$ (4) $4x - 5y = 10$

In each case find the intercepts on the x-axis and on the y-axis.

6 Solve graphically the following pairs of equations and check by algebraical solutions:

(1) $x + y = 7$ (2) $2x + y = 7$
    $x - y = 1$ $3x - 5y = 4$
(3) $2x + 3y = 2$
    $4x + y = -6$

7 The straight line whose equation is $y = ax - 1$ passes through the point $(2, 5)$. What is the value of a? What is the intercept on the y-axis?

8 The straight line whose equation is $y = 2x + b$ passes through the point $(1, 3)$. What is the value of b? Draw the straight line. What is its intercept on the y-axis?

9 The points $(1, 1)$ and $(2, 4)$ lie on the straight line whose equation is $y = ax + b$. Find a and b and write down the equation. What is the intercept of the straight line on the y-axis?

(*Hint:* See example 2 of Section 65.)

10 Draw the straight line $y = 2x + 1$.
Draw the straight line parallel to it and passing through the point $(0, 3)$. What is the equation of this line?

11 Find the equation of the straight line passing through the points $(1, 3)$ and $\left(2, \dfrac{11}{3}\right)$. What is the gradient of the line?

12 Find the equation of the straight line passing through the points (2, 4) and (3, 5). Find its gradient.

12

Using Inequalities to Define Regions

86 DEFINING REGIONS

In Fig. 38 the line $x = 3$ and the line $y = 4$ have been marked.

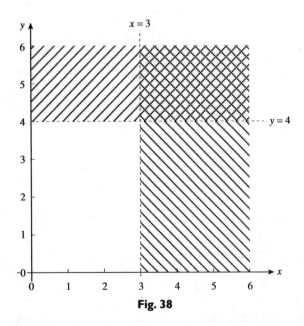

Fig. 38

All the points whose x-coordinate is greater than 3, shaded ⟍⟍, lie to the right of the line $x = 3$, whilst all the points whose coordinate is less than 3 lie to the left of the line $x = 3$.

Similarly all the points whose y-coordinate is greater than 4, shaded `⫽` , lie above the line $y = 4$, whilst all the points whose y-coordinate is less than 4 lie below the line $y = 4$.

The region which is shaded `XX:` is defined by the inequalities $x > 3$ and $y > 4$.

The region which is unshaded is defined by the inequalities $x < 3$ and $y < 4$.

The other two regions on this grid can be defined in a similar way.

87 REGIONS ABOVE AND BELOW STRAIGHT LINES

In Chapter 11 you saw how any equation of the form $y = mx + c$ could be represented by a straight line with a gradient of m and an intercept on the y-axis of c.

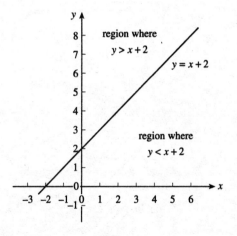

Fig. 39

This straight line, if extended indefinitely in each direction, divides the page into two regions, one above the line and the other below the line.

The coordinates of the points in the region above the line $y = x + 2$ are described by the inequality $y > x + 2$, whilst the coordinates of the points in the region below the line are described by $y < x + 2$. The coordinates of the points on the line itself satisfy $y = x + 2$. See Fig. 39.

■ *Example 1:* Show on a graph the region described by the inequalities $y > x$ and $y < 2x - 3$.

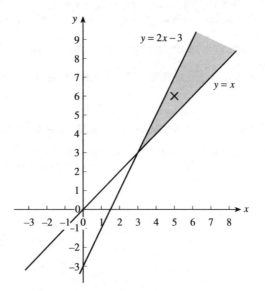

Fig. 40

The points in the region $y < 2x - 3$ lie *below* the line $y = 2x - 3$.
The points in the region $y > x$ lie *above* the line $y = x$.
The required region is shaded in the top right-hand part of the diagram, shown in Fig. 40.

Note that the point (5, 6) lies in this region. Check: $6 > 5$ and $6 < 2 \times 5 - 3$.

■ *Example 2:* Show on a graph and find one point in the region described by the inequalities $y > x - 2$, $y < 2x + 1$ and $x + y < 6$.

The points in the region $y > x - 2$ lie *above* the line $y = x - 2$.
The points in the region $y < 2x + 1$ lie *below* the line $y = 2x + 1$.
The points in the region $x + y < 6$ lie *below* the line $x + y = 6$.
The required region is shaded in the middle of the diagram, shown in Fig. 41.
One point in this region is (3, 2). Check: $2 > 3 - 2$, $2 < 2 \times 3 + 1$ and $3 + 2 < 6$.

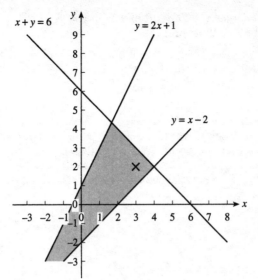

Fig. 41

EXERCISE 31

1 Draw the given line and then shade the required region. In each case find whether the point $(3, 1)$ is in the region or not.

    (*a*) line $x + y = 7$, region $x + y < 7$
    (*b*) line $y = 2x + 1$, region $y > 2x + 1$
    (*c*) line $y = 3x - 7$, region $y < 3x - 7$
    (*d*) line $2x + 3 = 12$, region $2x + 3y > 12$

2 Draw the given line and then shade the required region.

    (*a*) line $y = 2x$, region $y > 2x$
    (*b*) line $3x + 2y = 24$, region $3x + 2y < 24$

Shade also the region where both $y > 2x$ and $3x + 2y < 24$. Find whether the points $(2, 5)$ and $(6, 1)$ are in this region or not.

3 Draw the given line and then shade the required region.

    (*a*) line $y = x + 23$, region $y < x + 2$
    (*b*) line $y = 3x - 4$, region $y < 3x - 4$

Shade also the region where both $y < x + 2$ and $y < 3x - 4$. Find whether the points $(3, 4)$ and $(4, 6)$ are in this region or not.

4   Draw the lines $x + 2y = 12$ and $2x + y = 10$.
Mark the region where:

(*a*) $x + 2y > 12$ and $2x + y < 10$
(*b*) $x + 2y < 12$ and $2x + y > 10$
(*c*) $x + 2y < 12$ and $2x + y < 10$
(*d*) $x + 2y > 12$ and $2x + y > 10$

5   Draw the lines $y = x + 3$, $y = 3x - 1$ and $x + y = 11$.
Shade the region where $y > x + 3$, $y < 3x - 1$ and $x + y < 11$.
Find whether the points $(3, 6)$ and $(3, 7)$ are in this region or
not.

6   Using Fig. 42 below say which region is described by:

(*a*) E   (*b*) D   (*c*) C

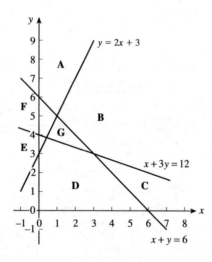

Fig. 42

7   Draw the lines $x + y = 7$, $y = 2x + 5$ and $x + 4y = 12$.
Mark the region where:

(*a*) $x + y \leq 7$, $y \leq 2x + 5$ and $x + 4y \geq 12$
(*b*) $x + y \geq 7$, $y \leq 2x + 5$ and $x + 4y \geq 12$
(*c*) $x + y \leq 7$, $y \geq 2x + 5$ and $x + 4y \geq 12$

8 Shade on a graph the region where:

    (a) $x + y < 10, 2x + y > 12$ and $x + 4y \geq 16$
    (b) $x + y < 10, 2x + y > 12$ and $x + 4y \geq 16$
    (c) $x + y > 10, 2x + y < 12$ and $x + 4y \leq 16$

88 GREATEST OR LEAST VALUES IN A REGION

■ *Example:* Shade on a graph the region where $x + 3y \leq 24$ and $3x + y < 21$. Find at what points in this region the expression $x + y$ takes on its largest value, with x and y being whole numbers.

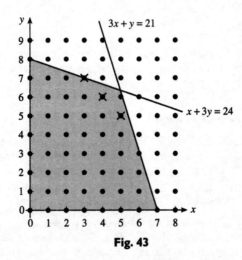

Fig. 43

The points in the shaded region in Fig. 43 for which $x + y$ is largest are (3, 7), (4, 6) and (5, 5). Points on the line $3x + y = 21$ are not included since $3x + y < 21$ and not $3x + y \leq 21$.

Note that the three points where $x + y$ is largest lie on the line $x + y = 10$.

EXERCISE 32

1 Look again at the previous graph.
Identify the region where $3x + y > 21$ and $x + 3y > 24$.
At which whole number point(s) is $x + y$ smallest in this region?

2 Shade on a graph the region where $2x + y \leq 10$ and $x + 2y < 12$.

Find at what point(s) in this region the expression $x + y$ takes on its largest value, with x and y being whole numbers.

3 Use your graph from Question 2 to identify the region where $2x + y > 10$ and $x + 2y > 12$.

Find at what point(s) in this region the expression $x + y$ takes on its smallest value, with x and y being whole numbers.

4 Shade on a graph the region where $y \geq 2x - 10$ and $x + 3y < 18$.

Find at what point(s) in this region, with whole number values for x and y, the expression $x + y$ takes on its largest value.

Explain why the point $(6, 4)$ is not included in the solution.

5 Use your graph from Question 4 to identify the region where $y < 2x - 10$ and $x + 3y \geq 18$.

Find at what point(s) in this region the expression $x + y$ takes on its smallest value, with x and y being whole numbers.

Explain why the point $(7, 4)$ is not included in this solution.

89 LINEAR PROGRAMMING

The ideas in Section 87 can be applied to a range of practical problems which occur quite frequently in business and industry. The constraints on a situation are described by a set of inequalities and then it is required, for example, to maximise the profit or supply, or to minimise the labour required, within these constraints.

■ *Example 1:* A manufacturer makes two types of smoke detector. For model A, the materials cost £4, the labour £5 and the profit is £5. For model B, the materials cost £3, the labour £9 and the profit is £6. Altogether £240 is available for the materials and £450 is available for the labour. How many of each model should be made to give the maximum profit?

If a is the number of model A and b is the number of model B to be made then you can show the constraints on materials and labour as a pair of inequalities: for materials $4a + 3b \leq 240$ and for labour $5a + 9b \leq 450$. You want to maximise the profit, which is given by £$(5a + 6b)$.

First show the two inequalities on a graph and find the region which corresponds to $4a + 3b \leq 240$ and $5a + 9b \leq 450$. In this example the region is the shaded region in Fig. 44 below each of the two lines.

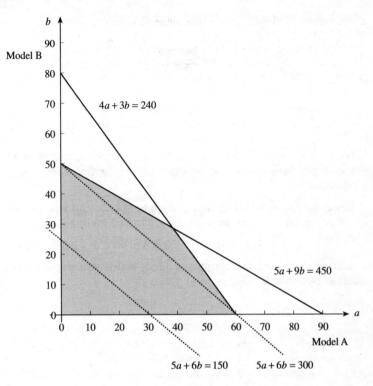

Fig. 44

You now need to find the point in this region where $5a + 6b$ is greatest. It would be very tedious to check the value of $5a + 6b$ at each point but fortunately you can use your knowledge of parallel lines to help. The lines $5a + 6b = 100$, $5a + 6b = 150$, $5a + 6b = 300$ etc. are all parallel. The second two are shown dotted on the graph in Fig. 44 and you should note that as the profit gets larger the corresponding profit line moves up and to the right. You need to find a point on the profit line which is furthest to the right in the required region. In this case the point is (39, 28) and the profit £363.

■ *Example 2:* A farmer makes a composite feed from two different feeds so that the vitamin content of the final feed is at least:

Vitamin A – 8 units, Vitamin B – 6 units, Vitamin C – 3 units.

The vitamin content of each feed per kilogram is:

	Vitamin A	Vitamin B	Vitamin C
Feed 1	1	3	1
Feed 2	4	1	1

If the costs of each feed per kilogram are 120 pence and 90 pence respectively, how should the farmer make his feed to keep costs to a minimum?

Let the number of kilograms in each feed be x and y respectively. Then the three constraints described by the inequalities are:

$$A: x + 4y \geq 8, \ B: 3x + y \geq 6 \text{ and } C: x + y \geq 3$$

The cost of making the composite feed, which you want to be as small as possible, is $(120x + 90y)$ pence

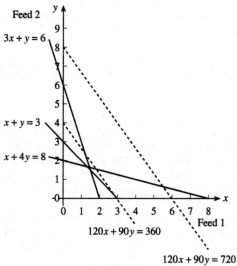

Fig. 45

On Fig. 45 the two cost lines for 360p and 720p are shown as dotted lines. These correspond to $120x + 90y = 360$ (i.e.

$4x + 3y = 12$) and $120x + 90y = 720$ (i.e. $4x + 3y = 24$). These lines are parallel and move down and to the left as the cost decreases. The minimum cost line will be the last line which contains at least one point in the required region.

In this example the minimum cost occurs when the composite feed consists of 1 part of Feed 1 to 3 parts of Feed 2.

EXERCISE 33

1 For a camp of 144 children, two types of tent are available for hire. The large tent sleeps 8 and costs £48 per week, and the small tent sleeps 3 and costs £12 per week.
The total number of tents must not exceed 36.
Using L for the number of large tents and S for the number of small tents write down inequalities to describe the constraints of:

    (*a*) the number of children   (*b*) the number of tents.

Write down an expression to show the cost of hiring these tents.
If the cost of hiring is to be kept as small as possible, show the information on a graph and find the number of each type of tent that must be hired.
What will be the total cost of hiring for one week?

2 A market gardener who has 100 hectares of land available for planting lettuces and/or spring onions is prepared to outlay at most £5,400. The initial outlay on each hectare of lettuces is £36, whilst that on each hectare of spring onions is £90.
Show this information as a pair of inequalities and represent it on a graph.
If the profit on each hectare of lettuces is £80 and on each hectare of spring onions is £120, find how the market gardener should allocate the land to make the maximum profit.
What is the greatest profit that could be made if 120 hectares was used?
How many hectares must be allocated to make it worth growing only lettuces?

3 A firm wants to buy a number of new cars of two types.
Type A costs £16,000 and £96 per week to run.
Type B costs £19,200 and £60 per week to run.
The firm has £144,000 to spend and wishes to keep its running costs to £720 per week.
If the firm must buy at least 4 cars of type A and 2 cars of type

B, find the various possibilities for the number of each type of car that the firm could buy.

4   A firm with 10 drivers uses vans and lorries to make its deliveries.

A van costs £72 per day to run and can carry 600 kg.

A lorry costs £180 per day to run and can carry 1100 kg.

The firm wants to keep its daily running costs down to £1,080, but at the same time wants to be able to deliver at least 6600 kg.

Write down inequalities to describe the constraints of:

  (*a*) drivers (*b*) running costs (*c*) carrying capacity.

Show this information on a graph and find the various possibilities for the number of vans and lorries if the firm decides to use at least 2 lorries.

5   Represent on one graph the set of points (x, y) which satisfy the inequalities: $x \geq 10$, $y \geq 2$, $x + 4y \geq 24$ and $2x + 3y \geq 36$.

Show also the region where $x + 2y \leq 30$ as well.

Hence find the points in this region, with whole number values for x and y, at which $x + 2y$ takes its smallest value.

13

Multiplication of Algebraical Expressions

90 (1) WHEN ONE FACTOR CONSISTS OF ONE TERM

This has been considered in Section 27, when it was shown that

$$x(a + b) = xa + xb.$$

91 (2) PRODUCT OF BINOMIAL EXPRESSIONS – I.E., WITH TWO TERMS

A typical example is $(x + a)(y + b)$.

As in Section 27, a geometrical illustration will help to make clear what is the product and how it is obtained.

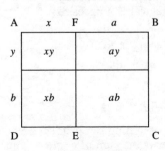

Fig. 46

In Fig. 46 ABCD is a rectangle with the sides $(x + a)$ and $(y + b)$ units of length, and divided to represent x, a, y, and b units of length. Lines are drawn parallel to the sides dividing the whole rectangle into smaller ones whose areas represent the products xy, xb, ay and ab by means of their areas.

The area of the whole rectangle

$$=(x + a)(y + b) \text{ sq. units.}$$

Also the area of the whole rectangle

$$= \text{area of ADEF} + \text{area of EFBC}$$

$$= x(y + b) + a(y + b).$$
$$\therefore \quad (x + a)(y + b) = x(y + b) + a(y + b)$$
$$= xy + bx + ay + ab.$$

This grouping suggests the method of multiplying $(x + a)(y + b)$ algebraically.

The second binomial $(y + b)$ is multiplied in turn by each term of the first factor. The sum of these is the final product.

As the order in multiplication is immaterial so far as the final product is concerned, this could also have been obtained as follows. Writing the factors in the reverse order:

$$(y + b)(x + a) = y(x + a) + b(x + a)$$
$$= xy + ay + bx + ab.$$

This is illustrated in Fig. 46.

Worked examples

$$(a + b)(c + d) = a(c + d) + b(c + d)$$
$$= ac + ad + bc + bd.$$
$$(x + 5)(y + 3) = x(y + 3) + 5(y + 3)$$
$$= xy + 3x + 5y + 15.$$
$$(a + b)(c - d) = a(c - d) + b(c - d)$$
$$= ac - ad + bc - bd.$$
$$(a - x)(b - y) = a(b - y) - x(b - y)$$
$$= ab - ay - bx + xy.$$
$$(x - 4)(y + 3) = x(y + 3) - 4(y + 3)$$
$$= xy + 3x - 4y - 12.$$

If the first terms in each factor are alike, the same method is followed, the product being simplified afterwards if necessary. Thus:

$$(x + a)(x + b) = x(x + b) + a(x + b)$$
$$= x^2 + bx + ax + ab.$$

This could be expressed as

$$x^2 + (b + a)x + ab.$$

This result suggests a quick way of obtaining the product mentally. The coefficient of x in the answer is the sum of a and b. The last term is their product.

When a and b are numbers the sum and product will be evaluated, and the expression simplified.

■ *Examples*

$$(x + 6)(x + 5) = x(x + 5) + 6(x + 5)$$
$$= x^2 + 5x + 6x + 30$$
$$= x^2 + 11x + 30.$$
$$(a + 9)(a + 4) = a^2 + (9 + 4)a + 9 \times 4$$
$$= a^2 + 13a + 36.$$
$$(x + 2)(x - 7) = x(x - 7) + 2(x - 7)$$
$$= x^2 - 7x + 2x - 14$$
$$= x^2 + x(-7 + 2) - 14$$
$$= x^2 - 5x - 14.$$
$$(a - 8)(a - 3) = a(a - 3) - 8(a - 3)$$
$$= a^2 - 3a - 8a + 24$$
$$= a^2 - 11a + 24.$$
$$(x - 8)(x - 2) = x^2 + (-8 - 2)x + (-8) \times (-2)$$
$$= x^2 - 10x + 16.$$
$$(x - 8)(x + 2) = x^2 + (-8 + 2)x + (-8 \times 2)$$
$$= x^2 - 6x - 16.$$
$$(x + 8)(x - 2) = x^2 + (+8 - 2)x + \{+8 \times (-2)\}$$
$$= x^2 + 6x - 16.$$

92 WHEN THE COEFFICIENTS OF THE FIRST TERMS ARE NOT UNITY

The rule still holds. Thus:

$$(px + a)(qx + b) = px(qx + b) + a(qx + b)$$
$$= pqx^2 + pbx + aqx + ab,$$

which you can write as

$$pqx^2 + (pb + aq)x + ab.$$

Note the form of the coefficients of x in the last line. It will be used later.

Numerical examples of this form are common. The following illustrations show how to apply the rule quickly.

Worked examples

■ *Example 1:* $(2x + 5)(3x + 4)$

$$= 2x(3x + 4) + 5(3x + 4)$$
$$= 6x^2 + x(2 \times 4 + 5 \times 3) + 20$$
$$= 6x^2 + 23x + 20.$$

You can obtain the coefficient of the middle term by multiplying as shown by the arrows below; then add the results:

$$(2x + 5)(3x + 4).$$

■ *Example 2:*

$$(3x + 7)(2x + 1).$$

Product $= 6x^2 + x\{(3 \times 1) + (7 \times 2)\} + 7$
$\qquad\quad = 6x^2 + 17x + 7.$

■ *Example 3:*

$$(7x - 5)(2x + 3).$$

Product $= 14x^2 + x\{(7 \times 3) - (5 \times 2)\} - 15$
$\qquad\quad = 14x^2 + 11x - 15.$

■ *Example 4:*

$$(3x - 2)(4x - 7).$$

Product $= 12x^2 + x\{(3 \times -7) + (-2 \times 4)\} + 14$
$\qquad\quad = 12x^2 - 29x + 14.$

If you find any difficulty set out the working as shown in the second line of Example 1.

93 MULTIPLICATION OF A TRINOMIAL

You can adapt the method shown in Section 91 to certain cases in which one of the factors is a trinomial.

■ *Example 1:* $(x + 2)(x^2 - x + 1)$

$\qquad = x(x^2 - x + 1) + 2(x^2 - x + 1)$
$\qquad = x^3 - x^2 + x + 2x^2 - 2x + 2$
$\qquad = x^3 + x^2 - x + 2$ (on collecting like terms).

■ ***Example 2:*** $(a + b)(a^2 - ab + b^2)$

$$= a(a^2 - ab + b^2) + b(a^2 - ab + b^2)$$
$$= a^3 - a^2b + ab^2 + a^2b - ab^2 + b^3$$
$$= a^3 + b^3 \text{ (on collecting like terms).}$$

EXERCISE 34

Write down the following products:

1	$(a + x)(b + y)$	**2**	$(c + d)(e + f)$
3	$(ax + b)(cy + d)$	**4**	$(a - x)(b - y)$
5	$(x - y)(a - b)$	**6**	$(a - x)(b + y)$
7	$(a + x)(b - y)$	**8**	$(a + 2)(b + 3)$
9	$(a - 2)(b - 3)$	**10**	$(a - 2)(b + 3)$
11	$(a + 2)(b - 3)$	**12**	$(x + 7)(x + 5)$
13	$(ab + 6)(ab + 3)$	**14**	$(x + 10)(x + 3)$
15	$(x - 10)(x - 3)$	**16**	$(x + 10)(x - 3)$
17	$(x - 10)(x + 3)$	**18**	$(p + 8)(p - 12)$
19	$(x - 4y)(x - 8y)$	**20**	$(x - 4y)(x + 8y)$
21	$(x + 4y)(x - 8y)$	**22**	$(a + 2b)(2a + 5b)$
23	$(3x - 4y)(3x - 5y)$	**24**	$(4x + 1)(7x + 2)$
25	$(2x - 3)(3x - 1)$	**26**	$(3x + 1)(3x - 4)$
27	$(1 + 3y)(1 - 4y)$	**28**	$(6x + 1)(3x - 5)$
29	$(7x - 3y)(2x + 5y)$	**30**	$(3a - 7b)(6a - 5b)$

31 Simplify $\{(x + y)(a + b) - (ay + yb)\}$ and divide the result by x.

32 Simplify $\{(a + b)(a - c) + bc\} \div a$.

33 Find the value of $(2x - y)(x + y) - (2x + y)(x - y)$ when $x = 3, y = 2$.

34 Simplify $(1 - 2x)(1 + 3y) - (1 - 2y)(1 + 3x)$ and find its value when $x = 0.1, y = 0.2$.

35 Find the following products:

    (a) $(x - y)(x^2 + xy + y^2)$
    (b) $(a + 2)(a^2 - 2a + 4)$
    (c) $(1 + x)(1 - x + x^2)$
    (d) $(x + a)(x^2 + 2ax + a^2)$

94 SQUARE OF A BINOMIAL EXPRESSION

In Section 91 it was shown that

$$(x + a)(x + b) = x^2 + x(a + b) + ab.$$

Since this is true whatever the values of the letters, let $b = a$.

Then
$$(x + a)(x + a) = x^2 + x(a + a) + a \times a$$
$$= x^2 + 2xa + a^2.$$

$\therefore$
$$(x + a)^2 = x^2 + 2ax + a^2.$$

If a is replaced by $-a$ throughout, then
$$(x - a)^2 = x^2 - 2ax + a^2.$$

Geometrical illustration.

Fig. 47, which is a modification of Fig. 46, shows an illustration of the above result by means of a square whose side is $x + a$.

Try to draw a similar figure to illustrate
$$(x - a)^2 = x^2 - 2ax + a^2.$$

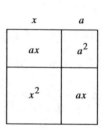

Fig. 47

■ *Examples*

$(x + 1)^2 = x^2 + 2x + 1$
$(x - 1)^2 = x^2 - 2x + 1$
$(a + 9)^2 = a^2 + 18a + 81$
$(1 - 5xy)^2 = 1 - 10xy + 25x^2y^2$
$(2x + 7y)^2 = 4x^2 + 28xy + 49y^2$
$(3a - 10b)^2 = 9a^2 - 60ab + 100b^2$

95 Square of a trinomial

In the product
$$(x + c)^2 = x^2 + 2cx + c^2.$$

Since x may have any value, replace it with $a + b$.
Then
$$(a + b + c)^2 = (a + b)^2 + 2c(a + b) + c^2$$
$$= a^2 + 2ab + b^2 + 2ac + 2bc + c^2,$$

or, re-arranging,
$$(a + b + c)^2 = a^2 + b^2 + c^2 + 2ab + 2bc + 2ac.$$

You can state this in words as follows:

The square of a trinomial is equal to
(sum of squares of each term) + (twice the product of the three terms taken two at a time in every possible way).

■ **Examples**

$$(a + b + 1)^2 = a^2 + b^2 + 1 + 2ab + 2a + 2b$$
$$(x - y + z)^2 = x^2 + (-y)^2 + z^2 + 2(x \times -y) + 2(xz) + 2(-y \times z)$$
$$= x^2 + y^2 + z^2 - 2xy + 2xz - 2yz$$
$$(a - b - c)^2 = a^2 + b^2 + c^2 - 2ab - 2ac + 2bc$$
$$(x + y - 5)^2 = x^2 + y^2 + 25 - 10x - 10y + 2xy$$

Take care in applying the laws of signs.

96 CUBE OF A BINOMIAL

$(a + b)^3$ may be written as $(a + b)(a + b)(a + b)$

$$= (a + b)(a^2 + 2ab + b^2).$$

Multiplying these as shown in Section 93

$$(a + b)^3 = a(a^2 + 2ab + b^2) + b(a^2 + 2ab + b^2)$$
$$= a^3 + 2a^2b + ab^2 + a^2b + 2ab^2 + b^3.$$

∴ $(a + b)^3 = a^3 + 3a^2b + 3ab^2 + b^3$ (on collecting like terms).

Similarly

$$(a - b)^3 = a^3 - 3a^2b + 3ab^2 - b^3.$$

■ **Examples**

$$(x + 1)^3 = x^3 + 3x^2 + 3x + 1$$
$$(x - 1)^3 = x^3 - 3x^2 + 3x - 1$$
$$(1 + a)^3 = 1 + 3a + 3a^2 + a^3$$
$$(1 - a)^3 = 1 - 3a + 3a^2 - a^3$$
$$(2x + 3y)^3 = (2x)^3 + \{3(2x)^2 \times 3y\} + \{3(2x) \times (3y)^2\} + (3y)^3$$
$$= 8x^3 + 36x^2y + 54xy^2 + 27y^3$$
$$(x - 3y)^3 = x^3 - 9x^2y + 27xy^2 - 27y^3$$

Note: In the cube of $(x - y)$ the signs are alternately positive and negative.

EXERCISE 35

Write down the following squares in full:

1 $(x + 2)^2$ **2** $(x - 2)^2$

3 $(a + 3b)^2$ **4** $(a - 3b)^2$

5 $(2x + y)^2$ **6** $(x - 2y)^2$

7 $(ab + 10)^2$ **8** $(xy - 3)^2$

9 $(4x + 5y)^2$ **10** $(4x - 5y)^2$

11 $(5xy + 6)^2$ **12** $(1 - 10x^2)^2$

13 $(5x^2 + 3y^2)^2$ **14** $(3xy - 2y^2)^2$

15 $\left(x + \dfrac{1}{y}\right)^2$ **16** $\left(\dfrac{1}{x} - \dfrac{1}{y}\right)^2$

17 $\left(a + \dfrac{1}{3}\right)^2$ **18** $\left(\dfrac{1}{2} - \dfrac{3y}{4}\right)^2$

19 $\{(x + y) + 1\}^2$ **20** $\{1 - (x - 2y)\}^2$

21 $(a + b - c)^2$ **22** $(x - y + z)^2$

23 $(2x + 3y - 5z)^2$ **24** $(4a - 2b - 1)^2$

Write down the following cubes in full:

25 $(x + y)^3$ **26** $(x - y)^3$

27 $(a + 2)^3$ **28** $(a - 2)^3$

29 $(p + q)^3$ **30** $(p - q)^3$

31 $(2x + y)^3$ **32** $(x - 2y)^3$

33 $(3a - 1)^3$ **34** $(1 - 3b)^3$

35   Simplify the following: $(x + y)^2 - (x^2 + y^2)$.

36   Simplify:

     (1) $(a + b)^2 - (a - b)^2$

     (2) $(a - b)^2 - (a + b)^2$

37   Simplify $(x + 10)^2 - (x - 10)^2$.

38   If $x = 3y + 1$, express $x^2 + 4x + 4$ in terms of y and find its value when $y = 1$.

39   A square lawn the side of which is x m is surrounded by a path (Fig. 48) which is a m wide. Find an expression for the area of the path in terms of x and a. What is the area of the path when $x = 30$ m and $a = 2$ m?

40   If the lawn of the previous question were a rectangle, x m by y m and the path were a m wide, find an expression for its area.

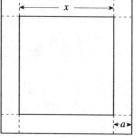

Fig. 48

97 PRODUCT OF SUM AND DIFFERENCE

If a and b are any two numbers, then the product of their sum and difference is expressed by $(a + b)(a - b)$.

Using the method of Section 91 to find the value of this

$$(a + b)(a - b) = a(a - b) + b(a - b)$$
$$= a^2 - ab + ab - b^2$$
$$\therefore \qquad (a + b)(a - b) = a^2 - b^2$$

This important result can be expressed in words as follows:

The product of the sum and difference of two numbers is equal to the difference of their squares.

■ *Examples*

$$(x + 9)(x - 9) = x^2 - 81$$
$$(ab + 10)(ab - 10) = a^2b^2 - 100$$
$$(4x + 5)(4x - 5) = 16x^2 - 25$$
$$(1 + x)(1 - x) = 1 - x^2$$
$$(a - \tfrac{1}{2})(a + \tfrac{1}{2}) = a^2 - \tfrac{1}{4}$$
$$\left(\frac{a}{3} + \frac{b}{4}\right)\left(\frac{a}{3} - \frac{b}{4}\right) = \frac{a^2}{9} - \frac{b^2}{16}$$
$$\{(a + b) + c\}\{(a + b) - c\} = (a + b)^2 - c^2$$
$$\{a + (b + c)\}\{a - (b + c)\} = a^2 - (b + c)^2$$

EXERCISE 36

Write down the following products:

1 $(a + x)(a - x)$ **2** $(p + q)(p - q)$

3 $(a + 2b)(a - 2b)$ **4** $(4x + 3)(4x - 3)$

5 $(2x + 1)(2x - 1)$ **6** $(1 + 6x)(1 - 6x)$

7 $(1 + a^2)(1 - a^2)$ **8** $(2x^2 + 1)(2x^2 - 1)$

9 $(x^2 + y^2)(x^2 - y^2)$ **10** $(3xy + 2)(3xy - 2)$

11 $(12xy + 1)(12xy - 1)$ **12** $(\tfrac{1}{4}x - 7)(\tfrac{1}{4}x + 7)$

13 $\{(x + y) + z\}\{(x + y) - z\}$ **14** $(a + x + y)(a + x - y)$

15 $(2a + 3b + 1)(2a + 3b - 1)$ **16** $(x - 2y + 6)(x - 2y - 6)$

17 $\{a + 2(b + c)\}\{a - 2(b + c)\}$

18 $\{2x + 3(y + z)\}\{2x - 3(y + z)\}$

19 $(x + \tfrac{2}{3})(x - \tfrac{2}{3})$

20 $\left(\dfrac{x}{2} + \dfrac{y}{3}\right)\left(\dfrac{x}{2} - \dfrac{y}{3}\right)$

14

Factors

98 The work of this chapter will be the converse of that in the previous one. That was concerned with methods of obtaining the products of certain algebraical expressions. This chapter is about finding the factors of expressions of different types.

A converse operation, in general, is more difficult than the direct one, and so it is in this case. By rules, on the whole simple and easily applied, the products of various kinds of factors are found. But in seeking to find the factors whose product produces a given expression, rules, even when they are formulated, are often long and tedious. In the main you have to rely on trial and improvement methods, which are, however, not haphazard but based on those rules by which the product was obtained.

It is always possible to obtain a product when the factors are given, but we cannot always find factors for expressions. Most expressions have no factors, so you can deal only with special types, such as were obtained as products in the work of the preceding chapter.

99 MONOMIAL FACTORS – i.e., FACTORS CONSISTING OF ONE TERM ONLY

This is the converse of the theorem stated in Section 27. There you saw that

$$x(a + b) = xa + xb.$$

If you start with $xa + xb$ and wish to factorise it, you see by inspection that x is a factor of each term. It is therefore a factor of the whole expression. To find the other factor you divide each term by x and add the quotients. As a result you get $a + b$.

$\therefore$ $\qquad\qquad xa + xb = x(a + b).$

In finding the factors by inspection you are guided by a knowledge of the process by which when $(a + b)$ is multiplied by x the result is $xa + xb$.

100 WORKED EXAMPLES

■ *Example 1:* Find the factors of $6a^2 + 3ac$.

In this case there is more than one factor common to each term.

(1) 3 is the highest common factor of the numerical coefficients.
(2) a is a factor of the other parts of each term.

$\therefore$ $3a$ is a factor of each term, and is therefore a factor of the whole expression. Dividing each term by it,

$$6a^2 + 3ac = 3a(2a + c).$$

■ *Example 2:* Factorise $5x^2y^2 - 10x^2y + 20y^2$.

The highest numerical factor common to each term is 5, and the only other factor is y.

$\therefore$ $\qquad 5x^2y^2 - 10x^2y + 20y^2 = 5y(x^2y - 2x^2 + 4y).$

EXERCISE 37

Express the following as factors:

1	$6x + 12$	**2**	$3ab + 2a$
3	$4xy + 2y^2$	**4**	$6a^2 - 4ab$
5	$14x^2y^2 - 7xy$	**6**	$16 - 32a^2$
7	$a^2 - ab + ac$	**8**	$x^3 + 3x^2 - x$
9	$15a^3 - 5a^2b + 3a^2b^2$	**10**	$6a^2c - 15ac^2$

11 $\quad a^2b + ab^2 - abc$ $\qquad$ **12** $\quad \dfrac{bc^2}{6} - \dfrac{abc}{9}$

13 Fill in the blank in the following:

$\qquad (7.4 \times 13^2) + (7.4 \times a^2) = 7.4($ $\qquad\qquad$).

14 Calculate the following as easily as you can:

$$18.6^2 + 18.6 \times 1.4.$$

101 BINOMIAL FACTORS

In Section 91 you saw that

$$(x + a)(y + b) = x(y + b) + a(y + b)$$
$$= xy + bx + ay + ab.$$

If you require the factors of $xy + bx + ay + ab$ and similar expressions, you must work backwards through the steps shown above.

The first step is to reach the stage $x(y + b) + a(y + b)$.

To obtain this the four terms of the expression must be suitably arranged in two pairs such that:

(1) the terms in each pair have a common factor;
(2) when this common factor is taken out, the same expression must be left in each pair.

∴ group $xy + bx + ay + ab$ in the form

$$(xy + bx) + (ay + ab)$$

Then taking out the common factor in each pair you get

$$x(y + b) + a(y + b).$$

Then $y + b$ is a factor of both parts of the expression and must therefore be a factor of the whole.

Thus you get, on taking it out,

$$(y + b)(x + a).$$

You can apply this method to expressions of which the factors are not previously known, as they were in the example.

102 WORKED EXAMPLES

■ *Example 1:* Find the factors of $a^2 + cd + ad + ac$.

The first two terms have no common factor. Consequently the order of the terms could be changed in order to get two pairs, each with a common factor.

∴ write

$$a^2 + cd + ad + ac = (a^2 + ad) + (cd + ac).$$

In this arrangement

(1) a is common to the terms of the first group and c to those of the second,

(2) the same expression $a + d$ will be the other factor of each group.

∴ you get $\qquad a(a + d) + c(a + d)$

$(a + d)$ being a factor of both parts is a factor of the whole.

∴ expression $\qquad = (a + d)(a + c).$

Note: Another possible arrangement was:

$$(a^2 + ac) + (ad + cd)$$
$$= a(a + c) + d(a + c)$$
$$= (a + c)(a + d).$$

There are always two possible ways of grouping.

■ *Example 2:* Factorise, if possible, $ab + ac + bc + bd$.

This example is given to show you where you might go wrong.

The expression can be grouped as $a(b + c) + b(c + d)$. But the expressions in the brackets in the two parts are different. It is a bad mistake, which is sometimes made, to write down the factors as $(b + c)(c + d)(a + b)$. Not one of these is a factor of the given expression. Moreover their product would be an expression of the third degree. The given expression is of the second degree.

On trying different groupings, you will find that you cannot arrange the expression in two groups, having the same factor in each group.

There are no factors of this expression.

■ *Example 3:* Find the factors of $ab - 5a - 3b + 15$.

By arrangement into suitable pairs:

$$ab - 5a - 3b + 15$$
$$= (ab - 5a) - (3b - 15)$$
$$= a(b - 5) - 3(b - 5)$$
$$= (b - 5)(a - 3).$$

Note carefully that when an expression is placed in brackets with minus sign in front, as was done with $-3b + 15$ on the previous page, the signs within the brackets must be changed. This is in agreement with the reverse rule given in Section 28, cases (3) and (4).

EXERCISE 38

Find the factors of:

1	$ax + ay + bx + by$	**2**	$pc + qc + pd + qd$
3	$ab - bd + ae - de$	**4**	$ax - cx - ay + cy$
5	$x^2 + px + qx + pq$	**6**	$x^2 - gx - hx + gh$
7	$ab + 5b + 6a + 30$	**8**	$ab - 5b - 6a + 30$
9	$ab - 5b + 6a - 30$	**10**	$2ab - 10a + 3b - 15$
11	$ax^2 + a^2x - ab - bx$	**12**	$x^2 + ax - bx - ab$

103　The form $x^2 + ax + b$

It was seen in Section 91 that the product of two factors such as $(x + 6)(x + 5)$ was $x^2 + 11x + 30$.

Reversing the process, consider how to find the factors of $x^2 + 11x + 30$.

In the general case it was shown that

$$(x + a)(x + b) = x^2 + x(a + b) + ab.$$

In this product

(1) the coefficient of x is the sum of the numbers a and b,
(2) the term independent of x, i.e., ab, is the product of these numbers.

Consequently in finding the factors of an expression such as $x^2 + 11x + 30$, you must find by trial two factors of 30 whose sum is $+11$. They are $+5$ and $+6$.

∴ the factors are $(x + 5)(x + 6)$.

104　Worked examples

■ *Example 1:* Factorise $x^2 + 13x + 36$.

There are several pairs of factors of 36 – namely (1×36), (2×18), (3×12), (4×9) and (6×6).

Look for the pair whose sum is $+13$; this pair is $(9, 4)$.

$$\therefore \qquad x^2 + 13x + 36 = (x + 4)(x + 9).$$

■ *Example 2:* Factorise $x^2 - 13x + 36$.

In this case the sign of the middle term is − and of the last term is +.

∴ look for two negative factors of 36 whose sum is −13.
These are −9 and −4.

$$\therefore \qquad x^2 - 13x + 36 = (x - 9)(x - 4).$$

■ *Example 3:* Factorise $y^2 - 13y + 30$.

Proceeding as in the last example, you get:

$$y^2 - 13y + 30 = (y - 10)(y - 3).$$

■ *Example 4:* Factorise $x^2 - 5x - 36$.

When the last term is negative the factors of 36 must have opposite signs.

∴ the coefficient of x is the sum of a positive and negative number, and the negative number must be the greater numerically, or, alternatively 5 is the difference of the two factors numerically.

The factors are −9 and +4.

$$\therefore \qquad x^2 - 5x - 36 = (x - 9)(x + 4).$$

Note: The larger of the two numbers has the same sign as the middle term.

■ *Example 5:* Factorise $x^2 + 12x - 28$.

The two factors of 28 which differ by 12 are 14 and 2. They are of opposite signs, and their sum is +12.

∴ the factors required are +14 and −2.

$$\therefore \qquad x^2 + 12x - 28 = (x + 14)(x - 2).$$

■ *Example 6:* Factorise $a^2 - 8ab - 48b^2$.

The introduction of the second letter makes no difference to the method followed, but b will appear in the second term of each factor.

Thus $\qquad a^2 - 8ab - 48b^2 = (a - 12b)(a + 4b).$

EXERCISE 39

Find the factors of:

1	$x^2 + 3x + 2$	**2**	$x^2 - 3x + 2$
3	$x^2 + 5x + 6$	**4**	$x^2 - 5x + 6$
5	$x^2 + 7x + 6$	**6**	$x^2 + 9x + 20$
7	$x^2 - 12xy + 20y^2$	**8**	$a^2 - 15ab + 36b^2$
9	$x^2y^2 + 15xy + 54$	**10**	$a^2b^2 - 19ab + 48$
11	$y^2 - 21y + 108$	**12**	$x^2 - 12xy + 35y^2$
13	$x^2 - x - 2$	**14**	$x^2 + x - 2$
15	$x^2 + xy - 6y^2$	**16**	$x^2 - xy - 6y^2$
17	$b^2 - 2b - 3$	**18**	$b^2 + 2b - 3$
19	$x^2 + 13x - 48$	**20**	$x^2 - 13x - 48$
21	$x^2 - xy - 110y^2$	**22**	$a^2 - 11a - 12$
23	$a^2 - a - 12$	**24**	$p^2 + p - 72$
25	$p^2 - 34p - 72$	**26**	$1 - 9x + 20x^2$
27	$1 - 8x - 20x^2$	**28**	$x^2y^2 - 3xy - 88$
29	$p^2 + 4p - 45$	**30**	$p^2 + pq - 56q^2$

105 THE FORM $ax^2 + bx + c$

The factors of an expression such as this, where a, b, c are numbers, will be of the form which occurred in the converse operation in Section 92. These factors are best obtained by trial, as indicated in the following examples.

■ *Example 1:* Find the factors of $2x^2 + 7x + 3$.

Write down possible pairs of factors systematically and find the middle term as shown in Section 92 until the correct coefficient of x is found. In this example the possibilities are:

(1) $(2x + 3)(x + 1)$.

(2) $(2x + 1)(x + 3)$.

In (1) the coefficient of x in the product is 5.
In (2) the coefficient of x in the product is 7.
The second pair is therefore the correct one – i.e.,

$$2x^2 + 7x + 3 = (2x + 1)(x + 3).$$

■ *Example 2:* Find the factors of $6x^2 + 17x - 3$.

The minus sign of the last term indicates that the factors will have opposite signs in the second terms. This increases the number of possible pairs of factors. Among the possible pairs are:

(1) $(2x - 1)(3x + 3)$ or $(2x + 1)(3x - 3)$
(2) $(6x - 1)(x + 3)$ or $(6x + 1)(x - 3)$

Proceeding as shown in Section 92, you find that the first pair of (2) is the one required, since the coefficient of x is

$$(6 \times 3) + (1 \times -1) = 18 - 1 = +17.$$
$$\therefore \qquad 6x^2 + 17x - 3 = (6x - 1)(x + 3).$$

■ *Example 3:* Find the factors of $4x^2 - 17x - 15$.

The sign of the last term is again negative; therefore the factors will have opposite signs. Among the possibilities are:

(1) $(2x + 5)(2x - 3)$ or $(2x - 5)(2x + 3)$
(2) $(4x + 5)(x - 3)$ or $(4x - 5)(x + 3)$
(3) $(4x + 3)(x - 5)$ or $(4x - 3)(x + 5)$

The first pair of (3) gives for the coefficient of x

$$(4 \times -5) + (3 \times 1) = -20 + 3 = -17.$$
$$\therefore \qquad 4x^2 - 17x - 15 = (4x + 3)(x - 5).$$

EXERCISE 40

Complete the following factors:

1 $3x^2 + 10x + 8 = (3x \qquad)(x \qquad)$
2 $12x^2 - 17x + 6 = (4x \qquad)(3x \qquad)$
3 $12x^2 - 28x - 5 = (6x \qquad)(2x \qquad)$
4 $9x^2 + 43x - 10 = (9x \qquad)(x \qquad)$

Find the factors of the following:

5	$2x^2 + 3x + 1$	6	$3x^2 - 4x + 1$
7	$2x^2 + 5x + 2$	8	$6x^2 + 5x + 1$
9	$4x^2 - 8x + 3$	10	$5x^2 - 6x + 1$
11	$6x^2 - 11x + 3$	12	$12x^2 + 11x + 2$
13	$2a^2 + a - 1$	14	$2a^2 - a - 1$
15	$2a^2 - a - 6$	16	$10b^2 - b - 2$
17	$10b^2 + b - 2$	18	$8y^2 - 14y - 15$
19	$12x^2 + 5x - 2$	20	$14c^2 - 17c - 6$

106 Squares of binomial expressions

You can recognise a trinomial expression which is the square of a binomial expression by applying the rule given in Section 94.

The standard forms are:

$$(a + b)^2 = a^2 + 2ab + b^2$$
and
$$(a - b)^2 = a^2 - 2ab + b^2.$$

If the trinomial is a square, the following conditions must be satisfied:

(1) the first and last terms, when the trinomial has been suitably arranged, are exact squares and positive;

(2) the middle term must be $\pm$ (twice the product of the square roots of the first and third terms).

■ *Example 1:* Write $x^2 + 6x + 9$ as a square of a binomial expression.

(1) The first and third terms are the squares of x and 3.
(2) The middle term is $+(2 \times \sqrt{x^2} \times \sqrt{9}) = 6x$.

$$\therefore \qquad x^2 + 6x + 9 = (x + 3)^2.$$

Similarly $\qquad x^2 - 6x + 9 = (x - 3)^2.$

■ *Example 2:* Is $4x^2 + 6x + 9$ a complete square?

(1) The first and third terms are the squares of $2x$ and 3.
(2) For a complete square the middle term should be $+(2 \times \sqrt{4x^2} \times \sqrt{9}) = +12x$.

But the middle term is $+6x$.

∴ the expression is not a complete square.

■ *Example 3:* Is $4x^2 - 20x + 25$ a complete square?

The middle term is

$$-2(\sqrt{4x^2}) \times \sqrt{25} = 2 \times 2x \times 5 = 20x.$$
$$\therefore \qquad 4x^2 - 20x + 25 = (2x - 5)^2.$$

107 Difference of two squares

In Section 97 it was shown that

$$(x + a)(x - a) = x^2 - a^2.$$

Conversely $\quad (x^2 - a^2) = (x + a)(x - a).$

∴ an expression which is the difference of two squares has for its factors the sum and the difference of the numbers which are squared.

Note: No real factors can be found for the sum of two squares – i.e., $x^2 + a^2$.

108 WORKED EXAMPLES

■ *Example 1:* Factorise $100x^2 - 1$.

The numbers squared are $10x$ and 1.
∴ the factors are the sum and difference of these – i.e.,

$$100x^2 - 1 = (10x + 1)(10x - 1).$$

■ *Example 2:* Factorise $36a^2b^2 - 25$.

The numbers squared are $6ab$ and 5.

$$\therefore \qquad 36a^2b^2 - 25 = (6ab + 5)(6ab - 5).$$

■ *Example 3:* Factorise $(a + b)^2 - c^2$.

Although one of the terms squared is replaced by a binomial expression, the rule above still applies.

$$\therefore \qquad (a + b)^2 - c^2 = \{(a + b) + c\}\{(a + b) - c\}$$
$$= (a + b + c)(a + b - c).$$

■ *Example 4:* Factorise $(a + b)^2 - (c - a)^2$.

$$(a + b)^2 - (c - a)^2 = \{(a + b) + (c - a)\}\{(a + b) - (c - a)\}$$
$$= (a + b + c - a)(a + b - c + a)$$
$$= (b + c)(2a + b - c).$$

109 EVALUATION OF FORMULAE

The difference of two squares formula is often used in transforming formulae and in arithmetical calculations.

■ *Example 1:* Find the value of $47.5^2 - 22.5^2$.

$$47.5^2 - 22.5^2 = (47.5 + 22.5)(47.5 - 22.5)$$
$$= 70 \times 25$$
$$= 1750.$$

■ ***Example 2:*** Find the area of a ring between two concentric circles of radii 97 mm and 83 mm, respectively. The area of a circle is πr^2 where the radius is r.

The area of the ring is the difference between the areas of the two circles.

$$
\begin{aligned}
\text{Difference in area} &= \pi \times 97^2 - \pi \times 83^2 \\
&= \pi(97^2 - 83^2) \text{ taking out common factor} \\
&= \pi\{(97 + 83) \times (97 - 83)\} \\
&= \pi \times 180 \times 14 \\
&= 2520\pi \text{ mm}^2.
\end{aligned}
$$

EXERCISE 41

Write each of the following as the square of a binomial expression:

1 $p^2 + 2pq + q^2$
2 $x^2 - 4xy + 4y^2$
3 $9x^2 + 6x + 1$
4 $16x^2 - 40xy + 25y^2$

5 $x^2 + x + \frac{1}{4}$
6 $\dfrac{a^2}{9} + \dfrac{ab}{3} + \dfrac{b^2}{4}$

Express as complete squares:

7 $(a + b)^2 + 4(a + b) + 4$
8 $(x - y)^2 - 10(x - y) + 25$

Find the factors of the following:

9 $x^2 - 100$
10 $a^2b^2 - 25$
11 $4x^2 - 9y^2$
12 $25a^2 - 16b^2$
13 $121x^2 - 36y^2$
14 $144p^2 - 169q^2$
15 $25 - 16a^2$
16 $1 - 225x^2$
17 $8a^2 - 50b^2$
18 $3x^2 - 75$
19 $5x^2 - 45y^2$
20 $(a + b)^2 - c^2$

21 $(x + 2y)^2 - 16z^2$
22 $1 - \dfrac{81}{16}y^2$

23 $x^2 - (y + z)^2$
24 $a^2 - (x - 2y)^2$
25 $(x - 8)^2 - 49$
26 $(a + b)^2 - (a - b)^2$

Find the numerical values of the following:

27 $65^2 - 35^2$
28 $82^2 - 68^2$
29 $49^2 - 39^2$
30 $24^2 - 18^2$
31 $4.25^2 - 1.75^2$
32 $17.5^2 - 12.5^2$

33 Find the value of $\pi(r_1^2 - r_2^2)$ when $\pi = \dfrac{22}{7}$, $r_1 = 12.5$, $r_2 = 8.5$.

34 If $S = \dfrac{v^2 - u^2}{2f}$ find S when $v = 14.5$, $u = 2.5$ and $f = 1.5$.

35 In the formula $w = k(D^2 - d^2)$ find w, when $k = 2.4$, $D = 8.5$, $d = 7.5$.

36 If $y = \dfrac{m(v^2 - u^2)}{2p}$ find y when $m = 24$, $v = 44$, $u = 16$ and $p = 32$.

37 If $s = \frac{1}{2}gt^2$ find the difference between the values of s when $t = 8$ and $t = 6$. Take $g = 10$.

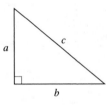

Fig. 49

38 In the right-angled triangle (Fig. 49) a and b represent the lengths of the sides containing the right angle and c the length of the hypotenuse. We know from geometry that

$$c^2 = a^2 + b^2.$$

(1) Find a when $c = 52$, $b = 48$.
(2) Find b when $c = 65$, $a = 25$.

110 SUM AND DIFFERENCES OF TWO CUBES

In Section 93, Example 2, you saw that

$$(a + b)(a^2 - ab + b^2) = a^3 + b^3.$$

By using the same method you get:

$$(a - b)(a^2 + ab + b^2) = a^3 - b^3.$$

These results enable you to obtain factors of the expressions:

$$a^3 + b^3 \text{ and } a^3 - b^3.$$

Re-arranging them you obtain:

$$a^3 + b^3 = (a + b)(a^2 - ab + b^2) \tag{A}$$
$$a^3 - b^3 = (a - b)(a^2 + ab + b^2) \tag{B}$$

It may help you to note that:

(1) in (A), the sum of two cubes, the first factor is the sum of the numbers;
(2) in (B), the difference of two cubes, the first factor is the difference of the numbers.

The other factors differ only in the sign of *ab*.

(1) When the first factor is the sum the sign is minus.
(2) When the first factor is the difference the sign is plus.

III WORKED EXAMPLES

$$x^3 + 1 = (x + 1)(x^2 - x + 1)$$
$$x^3 - 1 = (x - 1)(x^2 + x + 1)$$
$$1 - x^3 = (1 - x)(1 + x + x^2)$$
$$a^3 - 27 = (a - 3)(a^2 + 3a + 9)$$
$$8x^3 + 125 = (2x + 5)(4x^2 - 10x + 25)$$

EXERCISE 42

Find the factors of:

1 $x^3 + c^3$ **2** $y^3 - a^3$

3 $1 + 8a^3$ **4** $x^3 - 64$

5 $8 + 27c^3$ **6** $R^3 - 1$

7 $m^3 - 125n^3$ **8** $x^3y^3 + \frac{1}{8}$

9 $\dfrac{1}{x^3} + \dfrac{1}{y^3}$ **10** $\dfrac{1}{x^3} - \dfrac{1}{y^3}$

15

Fractions

112 ALGEBRAIC FRACTIONS

In algebra, fractions have the same fundamental meaning as in arithmetic, and are subject to the same rules in operations with them. They differ only in the use of letters in the algebraic forms.

An algebraic fraction is one in which the denominator is an algebraic expression. The numerator may or may not be algebraic in form.

Thus $\frac{2}{a}$ is an algebraic fraction, but $\frac{a}{2}$ is not. The latter means $\frac{1}{2}a$, and thus $\frac{1}{2}$ is merely a fractional coefficient.

Similarly $\frac{a}{a+b}$ is an algebraic fraction but $\frac{a+b}{6}$ is not.

113 LAWS OF FRACTIONS

It was stated above that algebraic fractions are subject to the same laws as arithmetical fractions. It is unnecessary to state these laws here, but a few illustrations may serve to remind you how they operate. It was pointed out in Section 26 that

$$\frac{a}{b} = \frac{a \times n}{b \times n} \quad \text{and} \quad \frac{a}{b} = \frac{a \div m}{b \div m} \, ,$$

where m and n are any numbers.

By means of this rule algebraic fractions can be simplified or reduced to lowest terms by dividing both numerator and denominator by the same number. This can be expressed by the phrase 'common factors are cancelled'. This was assumed in Section 23.

114 REDUCTION OF FRACTIONS

Examples were given in Section 26, but further instances are now given depending on the operations of the previous chapter.

■ ***Example 1:*** Simplify $\dfrac{a+b}{a^2-b^2}$.

Using the rule of Section 107.

$$\frac{a+b}{a^2-b^2} = \frac{a+b}{(a+b)(a-b)}$$

$$= \frac{1}{a-b} \quad \text{(on cancelling}$$

$(a+b)$).

■ ***Example 2:*** Simplify $\dfrac{x^2+4x-12}{x^2+x-6}$.

Factorising the numerator and denominator,

$$\frac{x^2+4x-12}{x^2+x-6} = \frac{(x-2)(x+6)}{(x-2)(x+3)}$$

$$= \frac{x+6}{x+3} .$$

The mistake is sometimes made of cancelling terms instead of factors. Thus, there can be no cancelling with the 3 and 6 in the above answer. Only factors of the whole expressions in numerator and denominator can be cancelled.

■ ***Example 3:*** Simplify $\dfrac{3a(a^2-4ab+4b^2)}{6a^2(a^2+3ab-10b^2)}$.

Factorising, the fraction becomes

$$\frac{3a(a-2b)^2}{6a^2(a+5b)(a-2b)} .$$

The factors $3a(a-2b)$ are common to numerator and denominator. Cancelling them, the fraction becomes,

$$\frac{a-2b}{2a(a+5b)} .$$

EXERCISE 43

Simplify the following fractions:

1 $\dfrac{2xyz}{6x^2z^2}$

2 $\dfrac{15abd^3}{20a^3bd}$

3 $\dfrac{2x^2 + 2xy}{4xy - 4y^2}$

4 $\dfrac{3a^3 + 9ab}{6a^3b - 6a^2b^2}$

5 $\dfrac{6x^2y + 6xy^2}{4x^3y^2 + 4x^2y^3}$

6 $\dfrac{a^2 - b^2}{a^2 - ab}$

7 $\dfrac{2x - 6}{x^2 - 5x + 6}$

8 $\dfrac{x^2 - 9x + 20}{x^2 + x - 20}$

9 $\dfrac{a^3 + 6a^2b + 9ab^2}{a^2b + 5ab^2 + 6b^3}$

10 $\dfrac{x^3 - xy^2}{x^4 - xy^3}$

115 MULTIPLICATION AND DIVISION

If necessary you should revise Section 26 and then proceed to the harder examples now given.

■ ***Example 1:*** Simplify $\dfrac{x}{x+1} \div \dfrac{x^2}{x^2 - 1}$.

Factorising and inverting the second fraction:

$$\frac{x}{x+1} \div \frac{x^2}{x^2 - 1} = \frac{x}{x+1} \times \frac{(x+1)(x-1)}{x^2}.$$

Cancelling common factors you get:

$$\frac{x-1}{x}.$$

■ ***Example 2:*** Simplify $\dfrac{x^4 - 27x}{x^2 - 9} \div \dfrac{x^2 + 3x + 9}{x + 3}$.

Factorising the expression is equal to:

$$\frac{x(x^3 - 27)}{x^2 - 9} \div \frac{x^2 + 3x + 9}{x + 3}$$

$$= \frac{x(x - 3)(x^2 + 3x + 9)}{(x + 3)(x - 3)} \times \frac{x + 3}{x^2 + 3x + 9}$$

$$= x \quad \text{(on cancelling factors).}$$

EXERCISE 44

Simplify the following fractions:

1 $\dfrac{xy}{x^2 - 1} \times \dfrac{x^2 - x}{y}$ **2** $xy \div \dfrac{x}{y + 1}$

3 $\dfrac{a^2 - b^2}{a^2 + 2ab + b^2} \times \dfrac{ab + b^2}{a^2 - ab}$ **4** $\dfrac{a^2 - 49}{a^2 - 9} \div \dfrac{a + 7}{a + 3}$

5 $\dfrac{2a^3 - 6a^2b}{2a^3 - 2ab^2} \times \dfrac{a^2 + 3ab + 2b^2}{a - 3b}$

6 $\dfrac{2x^2 + x - 1}{x - 2} \div \dfrac{2x^2 - x}{x^2 - 5x + 6}$ **7** $\dfrac{2a^2 - 8}{a + 2} \div 2a - 4$

8 $\dfrac{6x^2 - x - 2}{8x + 4} \times \dfrac{2}{3x^2 - 5x + 2}$

9 Simplify $\dfrac{2b}{b^2 - 1} \times \dfrac{b^2 - 3b - 4}{b^2}$ and find its value when $b = 5$.

10 Simplify $\dfrac{x^2 + 6x + 9}{2x - 2} \times \dfrac{x^2 - 1}{x^2 + 4x + 3}$ and find its value when $x = 1.5$.

116 ADDITION AND SUBTRACTION

The fundamental principles were examined in Section 25. Here are some more difficult examples.

■ *Example 1:* Simplify $\dfrac{a}{a - b} - \dfrac{a^2}{a^2 - b^2}$.

Factorising, the expression becomes

$$\dfrac{a}{a - b} - \dfrac{a^2}{(a + b)(a - b)} .$$

The least common denominator is $(a + b)(a - b)$.

$$\therefore \text{fraction} = \dfrac{a(a + b) - a^2}{(a + b)(a - b)}$$

$$= \dfrac{a^2 + ab - a^2}{(a + b)(a - b)}$$

$$= \dfrac{ab}{(a + b)(a - b)} .$$

■ **Example 2:** Simplify $\dfrac{3}{a-b} - \dfrac{2a+b}{a^2-b^2}$.

Factorising, the expression becomes

$$\dfrac{3}{a-b} - \dfrac{2a+b}{(a+b)(a-b)}$$

$$= \dfrac{3(a+b) - (2a+b)}{(a+b)(a-b)} = \dfrac{3a+3b-2a-b}{(a+b)(a-b)}$$

$$= \dfrac{a+2b}{(a+b)(a-b)}.$$

This example is intended to remind you of the note at the end of Section 28 relative to a minus sign before a fraction. The numerator, as $(2a + b)$ in the above example, should always be placed in a bracket on addition of the fraction, and the bracket removed afterwards.

■ **Example 3:** Simplify $\dfrac{x-1}{x^2-x-2} - \dfrac{x+2}{x^2+4x+3}$.

Factorising, the expression becomes

$$\dfrac{x-1}{(x-2)(x+1)} - \dfrac{x+2}{(x+3)(x+1)}$$

$$= \dfrac{(x-1)(x+3) - (x+2)(x-2)}{(x+1)(x-2)(x+3)}$$

$$= \dfrac{(x^2+2x-3) - (x^2-4)}{(x+1)(x-2)(x+3)}$$

$$= \dfrac{x^2+2x-3-x^2+4}{(x+1)(x-2)(x+3)}$$

$$= \dfrac{2x+1}{(x+1)(x-2)(x+3)}.$$

■ **Example 4:** Simplify $\dfrac{x}{x-\dfrac{1}{x}}$.

$$\dfrac{x}{x-\dfrac{1}{x}} = \dfrac{x}{\dfrac{x^2-1}{x}} = x \div \dfrac{x^2-1}{x}$$

$$= x \times \dfrac{x}{x^2-1} = \dfrac{x^2}{x^2-1}.$$

■ **Example 5:** Convert the formula $\dfrac{1}{R} = \dfrac{1}{R_1} - \dfrac{1}{R_2}$ into one in which the subject is R.

$$\frac{1}{R} = \frac{1}{R_1} - \frac{1}{R_2}$$

i.e., $\qquad \dfrac{1}{R} = \dfrac{R_2 - R_1}{R_1 R_2}.$

$\therefore \qquad\qquad R = \dfrac{R_1 R_2}{R_2 - R_1} \quad \left(\text{just as, if } \dfrac{1}{R} = \dfrac{1}{4}, \therefore R = 4\right).$

Note: You should not make the mistake of inverting at the outset. The fractions on the right side must be added first.

EXERCISE 45

Simplify the following:

1 $\dfrac{1}{x} + \dfrac{1}{x - y}$

2 $\dfrac{2}{x - 1} - \dfrac{3}{x - 2}$

3 $\dfrac{2}{3(a - b)} + \dfrac{3}{2(a + b)}$

4 $1 - \dfrac{1}{x} + \dfrac{1}{x + 2}$

5 $3 - \dfrac{2}{x - 3} + \dfrac{1}{(x + 3)^2}$

6 $\dfrac{2x}{3(x + 1)} - \dfrac{3x}{2(x - 2)}$

7 $\dfrac{3}{1 - a} + \dfrac{4}{(1 - a)^2}$

8 $\dfrac{x + 3}{x - 3} - \dfrac{x - 3}{x + 3}$

9 $\dfrac{1}{x} - \dfrac{1}{x + 2y} - 1$

10 $\dfrac{b}{3a - b} - \dfrac{a}{3b} + 1$

11 $\dfrac{1}{1 - at} - \dfrac{1}{1 - bt}$

12 $\dfrac{1}{a + bt} - \dfrac{1}{x + yt}$

13 $\dfrac{1}{x + y} + \dfrac{1}{x - y} - \dfrac{1}{x^2 - y^2}$

14 $\dfrac{x}{x^2 - y^2} - \dfrac{y}{(x - y)^2} - \dfrac{1}{x + y}$

Transform the following formulae so that the subject is R:

15 $\dfrac{1}{R} = \dfrac{1}{p} + \dfrac{1}{q}$

16 $\dfrac{1}{R} = \dfrac{1}{p} - \dfrac{1}{q}$

17 $\dfrac{1}{R} = \dfrac{2}{R_1} + \dfrac{3}{R_2}$ **18** $\dfrac{1}{R} = \dfrac{3}{R_1} - \dfrac{2}{R_2}$

19 If $\dfrac{1}{R} = \dfrac{1}{r+s} - \dfrac{1}{r-s}$ find R in terms of the other letters.

20 If $\dfrac{1}{P} = \dfrac{2}{p-q} + \dfrac{3}{p+q}$ find the value of $5P$.

21 Simplify $P + Q - \dfrac{P^2 + Q^2}{P + Q}$.

22 Simplify $\dfrac{P^2 + Q^2}{P - Q} - P$.

23 If $\dfrac{1}{R} = \dfrac{1}{R_1} + \dfrac{1}{R_2} + \dfrac{1}{R_3}$ find R when $R_1 = 8.6$, $R_2 = 4.3$, and $R_3 = 3$.

24 Given $a(P - \frac{1}{2}Q) = b(Q - \frac{1}{2}P)$, rearrange the terms so as to express P in terms of the other quantities.

25 Given $I = \dfrac{nE}{R + nr}$ find n in terms of the other quantities.

Find n when $I = 2$, $E = 1.8$, $R = 2.4$, $r = 0.5$.

117 Simple Equations Involving Algebraical Fractions

The following examples will illustrate the methods employed.

■ *Example 1:* Solve the equation:

$$\frac{3}{x - 2} = \frac{5}{x - 1}$$

Use the least common denominator, $(x - 2)(x - 1)$, to multiply both sides.

Then $\qquad \dfrac{3(x - 2)(x - 1)}{x - 2} = \dfrac{5(x - 2)(x - 1)}{x - 1}$

Cancelling $\qquad\qquad 3(x - 1) = 5(x - 2)$
or $\qquad\qquad\quad\; 3x - 3 = 5x - 10.$
$\therefore \qquad\qquad\qquad\quad 2x = 7$
and $\qquad\qquad\qquad\;\; x = 3.5.$

■ *Example 2:* Solve for n the equation:

$$\frac{1}{n - 2} + \frac{1}{n - 3} = \frac{2}{n}.$$

Multiply throughout by the least common denominator $n(n-2)(n-3)$.

Then $n(n-3) + n(n-2) = 2(n-2)(n-3)$.

∴ $n^2 - 3n + n^2 - 2n = 2(n^2 - 5n + 6)$

or $2n^2 - 5n = 2n^2 - 10n + 12$.

The terms involving $2n^2$ disappear.

∴ $5n = 12$

and $n = \dfrac{12}{5}$.

EXERCISE 46

Solve the following equations:

1 $\dfrac{18}{2x} - 1 = 2$ **2** $\dfrac{4x}{2x-1} = 5$

3 $\dfrac{x-1}{x-2} = 3$ **4** $\dfrac{1-x}{x+1} = 4$

5 $0.8 = \dfrac{1.5n}{4+1.4n}$ **6** $\dfrac{5}{2x+5} = \dfrac{4}{x+5}$

7 $\dfrac{3p+23}{3p+12} = \dfrac{4}{3}$ **8** $\dfrac{\dfrac{1}{x} - 3}{\dfrac{1}{x} - 2} = \dfrac{1}{4}$

9 $\dfrac{3}{x-2} - \dfrac{5}{x+1} = \dfrac{5}{(x-2)(x+1)}$

10 $\dfrac{3}{x} - \dfrac{2}{x-2} = 0$

11 In the formula $\dfrac{1}{f} = \dfrac{1}{v} - \dfrac{1}{u}$ find v when $f = 8$, $u = 2$.

12 For what value of n is $\dfrac{5n-20}{4n+6}$ equal to $\tfrac{1}{3}$?

13 If $C = \dfrac{V}{R}$ find R when $C = 7.5$, $V = 60$.

14 If $r = \dfrac{R(E-V)}{V}$ find V in terms of r, R, E.

16

Graphs of Quadratic Functions

118 CONSTANTS AND VARIABLES

In the formulae which have been considered in earlier chapters you have seen that the letters used represent two different kinds of numbers.

(1) Some represent constant numbers – i.e., numbers which remain unchanged in the varying cases to which the formula applies.
(2) Others represent variable quantities.

■ *Example 1:* Consider the formula for the circumference of a circle, namely.

$$C = 2\pi r,$$

where C = the length of the circumference,
               r = the length of the radius.

Two of these four symbols represent constants.

(1) The number 2 (which is fixed for any circle),
(2) The number π (which has always the same value in any formula in which it is used). It represents the constant ratio of the circumference of any circle to its diameter, or the area of a circle to the square of its radius.

Two of the letters in the formula are variables, they are different in value for different circles, the variation of C depending on changes in r.

■ *Example 2:* In Section 78 it was shown that a formula, or equation connecting profit and customers, was

$$y = 5x - 115$$

It was pointed out that while the profit represented by y depended on the varying number of customers x, the numbers 0.05 and 11.5 remained constant, representing respectively the average amount paid by each customer, and the fixed charges.

■ *Example 3:* This last example was a special case of the equation of a straight line, which in Section 84 was shown to be represented in general by the equation:

$$y = mx + b.$$

For a particular straight line, m is constant, representing the gradient of the line, while b is the fixed distance intercepted on the y-axis. But x and y vary for different points on the line. They represent the coordinates of any point.

119 DEPENDENT AND INDEPENDENT VARIABLES

The variables are seen to be of two kinds.

Considering the case of the circumference of the circle, the length of this depends on the length of the radius.

In the second example the profit depends on the number of customers.

A variable which thus depends on another variable for its value is called a **dependent variable**.

The other variable upon which the first one depends is called the **independent variable**.

As another example, if a train is moving with uniform – i.e. constant – speed, the distance travelled depends on the time.

Thus the speed is a constant,

> time is an independent variable,
> distance is a dependent variable.

Again the cost of a quantity of tea depends on the weight bought, the price per kilogramme being constant.

Thus   weight is the independent variable,
               cost is the dependent variable.

Graphs. In plotting graphs which show how one quantity varies as another varies, the independent variable is always measured on the x-axis and the dependent on the y-axis.

120 Functions

When two quantities are related as shown in the following examples, the dependent variable is said to be a function of the independent variable. Thus:

The circumference of a circle is a function of its radius.

The area of a square is a function of its side.

The distance travelled by a car moving uniformly is a function of the time.

If a spring is stretched by a force, the extension of the spring is a function of the force.

Generally if a quantity denoted by y depends for its value on another quantity x then y is a function of x.

Thus in each of the following examples

$$y = 2x$$
$$y = x^2$$
$$y = \frac{5}{x} - 3$$
$$y = 3(x - 1)(x - 3)$$

the value of y depends on the value of x. If any particular value is given to x, a corresponding value of y can be calculated. In all such cases y is a function of x.

The idea of a function is probably the most important in modern mathematics. From the above examples the following definition can be deduced:

Function. If a quantity y is related to a quantity x so that for every value which might be assigned to x there is a corresponding value of y, then y is a function of x.

Thus for every length that may be chosen for the radius of a circle, there is always a corresponding length of the circumference.

If a person works at a fixed rate per hour, then for any number of hours worked, there is always a corresponding amount of pay. The pay is a function of the time worked; the rate per hour is a constant.

Again if $y = 2x^3$, then for every value which you may choose to give to x, there is a corresponding value of y. Consequently, y is a function of x.

121 Graph of a function

If y is a function of x, and since, by the above definition, for every value assigned to x there is always a corresponding value of y, then these pairs of values of x and y can be plotted, and the assemblage of points so plotted will be the graph of the function.

Thus every function has a distinctive graph, which will often be a continuous curve or a straight line, and by which it can be identified.

If the function is of the first degree, and does not involve any algebraical fraction, then, as you have already seen, the graph will be a straight line. For this reason a function of the first degree, of which the general form is $y = mx + b$, is called a linear function.

If, however, the function involves a higher power of x, such as x^2, x^3, etc., or involves an algebraic fraction such as $y = \dfrac{1}{x}$,

the graph will be a smooth curve, the shape of which will differ with the nature of the function.

122 GRAPH OF A FUNCTION OF SECOND DEGREE

The simplest form of a function of the second degree is that which is expressed by $y = x^2$. This is called a quadratic function, from the Latin quadratus (squared). The area of a circle, $A = \pi r^2$, is a special form of this.

To plot the curve of $y = x^2$, first assign values to x, then calculate the corresponding values of x^2, or y, and tabulate them as follows:

x	-3	-2.5	-2	-1.5	-1	0	1	1.5	2	2.25	3
y	9	6.25	4	2.25	1	0	1	2.25	4	6.25	9

Take as wide a range of values as the size of the paper will allow you to plot. Since the values of y increase more rapidly than those of x, more room is needed on the y-axis, but as the square of a number is always positive, no negative values of y are necessary. The x-axis is therefore drawn near the bottom of the paper, as shown in Fig. 50.

Selection of scales on axes.   The scale chosen must depend on the number of values you wish to include within the limits of the paper. You need not choose the same scales on the two axes, and in this particular curve it will be better to take a smaller scale for values of y, because their values increase more rapidly than those of x.

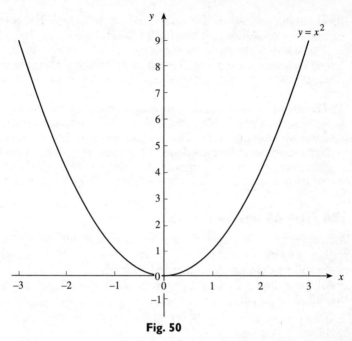

Fig. 50

You must remember, however, that the true shape of the curve will be shown only when you take the same units on both axes.

The symmetry of the values of y, as shown in the table, suggests that the y-axis should be in the middle of the paper.

The points when plotted appear to lie on a smooth curve as shown in Fig. 50.

The curve is called a parabola. When inverted it is the curve described by a projectile such as a shot from a gun, or a rocket when fired into the air ignoring air resistance.

123 SOME PROPERTIES OF THE GRAPH OF $y = x^2$

Note the following points about this curve.

(1) The symmetry of the curve. Positive and negative values of x produce equal values for y. If therefore the curve were folded about the y-axis, the two parts of the curve, point by point, would coincide.

The curve is therefore said to be symmetrical about the y-axis; the y-axis is called the **axis of symmetry.**

(2) The minimum value of the curve is 0 at the origin. The curve is said to have a turning-point at the origin.

(3) The slope of the curve is not constant, as with a straight line, but increases from point to point as x increases. The gradient is clearly a function of x, since its value depends on the value of x.

(4) The curve may be used to read off the square of any number within the range of plotted values, and also, conversely, to determine square roots. Thus to find $\sqrt{3}$, take the point on the curve corresponding to 3 on the y-axis. It is then seen that there are two points on the x-axis which correspond to this, the values of x (i.e., $\pm\sqrt{3}$) being $+1.73$ and -1.73 (see Section 45).

124 THE GRAPH OF $y = -x^2$

All values of y for this curve are equal numerically to the corresponding values of y in $y = x^2$, but are negative. The shape of the curve will be the same, but inverted, as in Fig. 51.

All the values of y being negative, the x-axis is drawn toward the top of the paper.

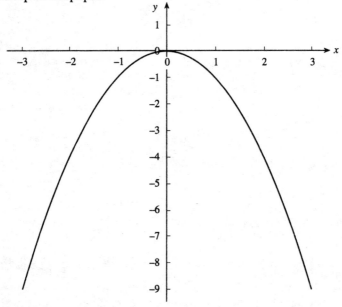

Fig. 51

The curve has a maximum point – namely zero – at the origin. From this point the curve shows the path of a bomb dropped from an aeroplane at 0 – ignoring air resistance.

125 THE GRAPHS OF $y = ax^2$

The curves represented by $y = ax^2$, where a is any number, are all parabolas differing from $y = x^2$ only in having different slopes.

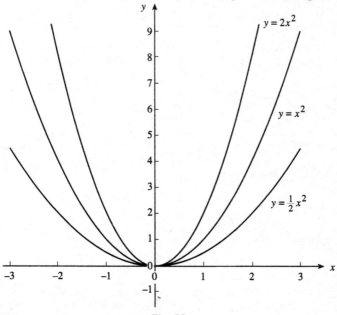

Fig. 52

Considering $y = 2x^2$ as an example the table of values for plotting the curve are found in the same way as for $y = x^2$.

This curve, as well as that of $y = \frac{1}{2}x^2$, are shown in Fig. 52, and contrasted with $y = x^2$.

If a is a negative number, you get a corresponding set of curves similar to $y = -x^2$, as in Fig. 51.

126 THE GRAPHS OF $y = x^2 \pm a$, WHERE a IS ANY NUMBER

The graph of $y = x^2 + 2$ is related to that of $y = x^2$ in the same way that $y = x + 2$ is connected with $y = x$ (see Section 84). Each ordinate of $x^2 + 2$ is greater by 2 than the corresponding ordinate of x^2.

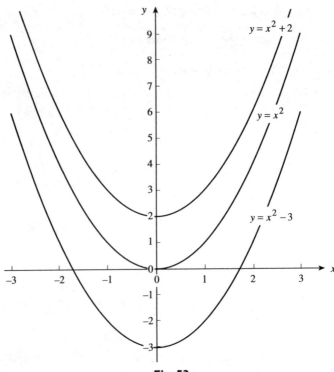

Fig. 53

The graph of $y = x^2 + 2$ is therefore the same as $y = x^2$ raised two units on the y-axis. It appears as shown in Fig. 53. Similarly, the curve $y = x^2 - 3$ is the graph of $y = x^2$, but every point is three units lower than $y - x^2$ for corresponding values of x.

Generally the curves represented by the equation $y = x^2 \pm a$ are a set of curves similar to $y = x^2$ and raised or lowered by an amount equal to $\pm a$, according to the sign of a.

Similarly, the graph of $y = -x^2$ gives rise to a set of curves included in the equation $y = -x^2 \pm a$.

127 GRAPH OF $y = (x - 1)^2$

The table below shows the values of (1) $y = (x - 1)^2$ and (2) $y = x^2$, for corresponding values of x.

x	-3	-2	-1	0	1	2	3	4
x^2	9	4	1	0	1	4	9	16
$(x - 1)^2$	16	9	4	1	0	1	4	9

A comparison of the sets of values of two functions shows that both have the same sequence of values, but those of $(x - 1)^2$ are those of x^2 moved one place to the right in the table for consecutive values of x.

Consequently the curve of $(x - 1)^2$ must be the same as that of x^2, but moved one unit to the right. See Fig. 54.

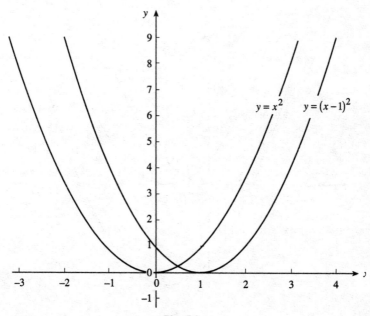

Fig. 54

Similarly, if you draw the graph of $(x - 2)^2$, you will find that it is the same as the graph of x^2, but moved two units to the right. Or if you draw the graph of $(x + 1)^2$, it is the same as the graph of x^2, but moved -1 units to the right, that is, 1 unit to the left.

128 GRAPH OF $y = (x - 1)^2 - 4$

By combining the operations illustrated in the previous two paragraphs – i.e., by moving the graphs – you can obtain the graph of a function such as $y = (x - 1)^2 - 4$ from the graph of $y = x^2$. You can do this in two steps.

(1) By moving the graph of $y = x^2$ one unit to the right to get the graph of $y = (x - 1)^2$, as in Section 127.
(2) By now moving the graph of $y = (x - 1)^2$ a distance of 4 units down the y-axis, the graph of $y = (x - 1)^2$ is transformed to $y = (x - 1)^2 - 4$.

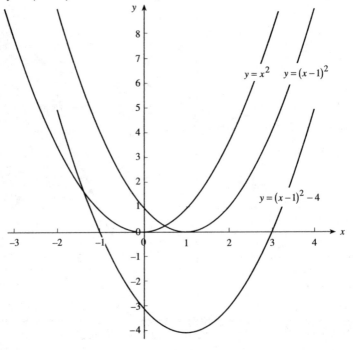

Fig. 55

These three graphs, $y = x^2$, $y = (x - 1)^2$ and $y = (x - 1)^2 - 4$ are shown in Fig. 55.

The expression $(x - 1)^2 - 4$ may be simplified, since

$$(x - 1)^2 - 4 = x^2 - 2x + 1 - 4$$
$$= x^2 - 2x - 3.$$

The curve lowest drawn in Fig. 55 is therefore the graph of

$$y = x^2 - 2x - 3.$$

You will see later (see Section 137) that you can reduce every quadratic function of x to such a form as this, or the simpler forms of Sections 126 and 127. It follows that:

The graph of every quadratic function of x is a modified form of $y = x^2$ and is therefore a parabola.

129 THE GRAPH $y = x^2 - 2x - 3$

You saw in the preceding paragraph that this graph can be obtained from the curve of $y = x^2$ by moving it parallel to the x- and y-axes. In practice, however, the method given below will be found more convenient in most cases and is usually more accurate in practical application. A table of values is constructed as follows:

x	-3	-2	-1	0	1	2	3	4
x^2	9	4	1	0	1	4	9	16
$-2x$	$+6$	$+4$	$+2$	0	-2	-4	-6	-8
-3	-3	-3	-3	-3	-3	-3	-3	-3
y	12	5	0	-3	-4	-3	0	5

Note: Be careful not to add the value of x in each column. It is well to draw a thick line under the values as shown, to remind you when adding.

From the values in this table you should plot the curve and compare the figure obtained with that in Fig. 55, arrived at by moving the graphs. It will be seen that the minimum value of $x^2 - 2x - 3$ is -4, when $x = 1$, and the line $x = 1$, perpendicular to the x-axis, is an axis of symmetry.

130 SOLUTION OF THE EQUATION $x^2 - 2x - 3 = 0$
FROM THE GRAPH

At the points where the graph of $x^2 - 2x - 3$, in Fig. 55, cuts the x-axis, the value of y is zero – i.e.:

$$x^2 - 2x - 3 = 0.$$

The values of x at these two points are:

$$x = 3 \text{ and } x = -1.$$

These are the two values of x which satisfy the equation $x^2 - 2x - 3 = 0$. You have therefore solved the equation by using the graph.

131 GRAPH OF $y = 2x^2 - 3x - 5$

This is a slightly more difficult graph to plot. The table of values is as follows:

x	-3	-2	-1	0	$\frac{1}{2}$	1	2	3	4	5
$2x^2$	18	8	2	0	$\frac{1}{2}$	2	8	18	32	50
$-3x$	9	6	3	0	$-1\frac{1}{2}$	-3	-6	-9	-12	-15
-5	-5	-5	-5	-5	-5	-5	-5	-5	-5	-5
y	22	9	0	-5	-6	-6	-3	4	15	30

In this example a wider range of values is shown. As the values of y increase rapidly smaller units are taken on the two axes. The graph is as shown in Fig. 56.

The lowest point, N, giving the minimum value of the expression, corresponds to $x = \frac{3}{4}$. From the table of values this point is seen to be half-way between $x = \frac{1}{2}$ and $x = 1$, since these values of x give the same values of y, namely -6. The minimum value, represented by MN, is $-6\frac{1}{8}$. The ordinate through M, of which MN is part, is the axis of symmetry of the curve.

The solution of the equation:

$$2x^2 - 3x - 5 = 0,$$

will be given by the values of x, where $y = 0$ – i.e., at A and B, where the curve cuts the x-axis. These points give:

$$x = -1, \text{ or } 2.5.$$

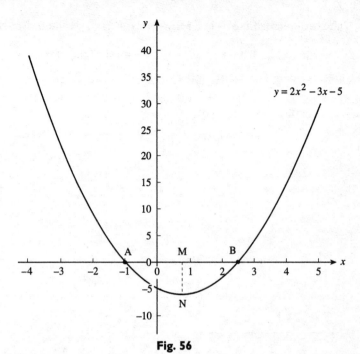

Fig. 56

132 GRAPH OF $y = 12 - x - x^2$

In this example the coefficient of x^2 is negative. Consequently the curve will take the form of $y = -x^2$, as shown in Fig. 51 – i.e., it will be an inverted parabola. Here is the table of values.

x	-5	-4	-3	-2	-1	0	1	2	3	4
12	12	12	12	12	12	12	12	12	12	12
$-x$	5	4	3	2	1	0	-1	-2	-3	-4
$-x^2$	-25	-16	-9	-4	-1	0	-1	-4	-9	-16
y	-8	0	6	10	12	12	10	6	0	-8

The values of y show the symmetry of the curve. The highest point, giving the maximum value, is seen to be half-way between $x = -1$, and $x = 0$ – i.e., where $x = -\frac{1}{2}$. The curve is as shown in Fig. 57. The maximum value is at M, where the value y is $12\frac{1}{4}$.

The ordinate from M – i.e., MN – is the axis of symmetry for the curve.

You can obtain from the curve the solution of the equation:

$$12 - x - x^2 = 0, \text{ i.e. } x^2 + x - 12 = 0.$$

This will be given by the values of x at A and B, where the curve cuts the x-axis. The solutions are:

$$x = 3, x = -4.$$

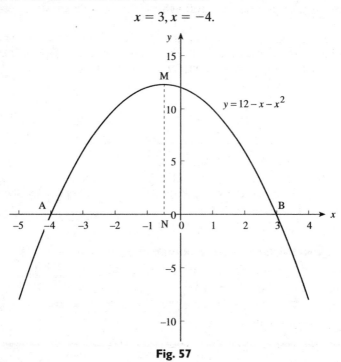

Fig. 57

133 USING GRAPHICS CALCULATORS

You can obtain all the graphs in the previous sections by using a graphics calculator. Different types and makes of graphics calculator vary slightly in the details of the method by which you obtain the graph, and you will need to read the manual to find the exact method for your calculator.

In all the different types of graphics calculator, you can specify the size of the window, which is the part of the graph which you wish to draw, by giving the minimum and maximum values of x and y for the axes. There are also features in most graphics calculators which enable you to move a flashing point along the curve and find its coordinates as you move it.

You are advised to draw some of the graphs in Exercise 47 'by hand' so that you understand more clearly what they mean. But if you have a graphics calculator, it is sensible to use it when appropriate in the exercises which follow in this and subsequent chapters.

In some cases you may find it best to draw the curves on your graphics calculator, and then to make a rough sketch copy of them on ordinary paper in order to answer the question.

EXERCISE 47

1 Draw the curve of $y = x^2$ between the values $x = +3$ and $x = -3$, taking the units as large as possible. From the curve write down the values of:

   (1) 2.3^2 (2) $\sqrt{7}$ (3) $\sqrt{3.5}$

2 Draw the curve of $y = \frac{1}{4}x^2$ between $x = +4$ and $x = -4$. Find from the curve the values of x such that:

   (1) $\frac{1}{4}x^2 = 0.8$ (2) $\frac{1}{4}x^2 = 2$ (3) $x^2 = 12$

3 Draw the curve of $-\frac{1}{2}x^2$ between $x = +4$ and $x = -4$. From the curve find the values of:

   (1) $-\frac{1}{2}x^2 = -3$ (2) $-x^2 = -5$ (3) $x^2 = 2$

4 Draw the curve of $y = x^2$ and hence obtain the curve of $y = x^2 + 3$. From the curve find the values of x such that:

   (1) $x^2 + 3 = 5$ (2) $x^2 + 3 = 9$

5 Draw the curve of $y = x^2$ and hence obtain the curves of:

   (a) $y = (x - 2)^2$ (b) $y = (x + 3)^2$

From the curves find the values of x such that:

   (1) $(x - 2)^2 = 3$ (2) $(x + 3)^2 = 1$
   (3) $(x - 2)^2 - 5 = 0$

6   Draw the curve of $y = (x + 2)^2 - 2$ by using the curve of $y = x^2$. Use the curve to find the values of x when:

   (1) $(x + 2)^2 - 2 = 0$ (2) $(x + 2)^2 - 1 = 0$
   (3) $(x + 2)^2 = 8$

7   Draw the curve of $y = x^2 - 6x + 5$. Find the least value of this function and the corresponding value of x. Use the curve to find the values of x when:

   (1) $x^2 - 6x + 5 = 0$ (2) $x^2 - 6x + 5 = 6$

8   Draw the curve of $y = x^2 - 4x + 2$. Find the minimum value of the function and the corresponding value of x. Use the curve to solve the following equations:

   (1) $x^2 - 4x - 2 = 0$ (2) $x^2 - 4x - 2 = 3$

9   Draw the curve of $y = 2x^2 - 5x + 2$. Find the minimum value of the function and the corresponding value of x. From the curve solve the equations:

   (1) $2x^2 - 5x + 2 = 0$ (2) $2x^2 - 5x - 1 = 0$

10   Draw the curve of $y = 2 - x - x^2$. Find the maximum value of the function and the corresponding value of x.

134 USING GRAPHS TO SOLVE QUADRATIC INEQUALITIES

Once you have drawn the graph of a quadratic function you can use it to write down the solutions for an associated quadratic inequality.

■ *Example:* By drawing the graph of $y = x^2 - 4x + 3$ find the values of x for which $x^2 - 4x + 3 > 0$.

   The values of x for which $x^2 - 4x + 3 > 0$ correspond to the places on the graph where $y > 0$. See Fig. 58. These are where y is positive or above the x-axis. The values of x where y is above the axis are where $x < 1$ or where $x > 3$.

   The values of x for which $x^2 - 4x + 3 < 0$ correspond to the places on the graph in Fig. 58 where $y < 0$. These are where y is negative or below the x-axis. The values of x, where y is below the axis, are between $x = 1$ and $x = 3$, i.e. $-1 < x < 3$.

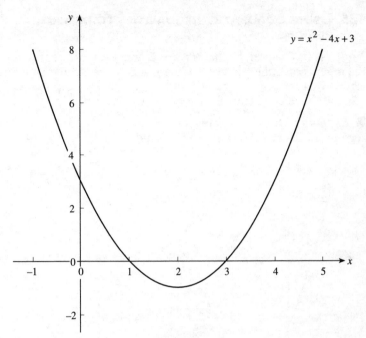

Fig. 58

EXERCISE 48

1 Draw the graph of $y = x^2 - 3x + 2$. Write down the values of x where the graph lies above the x-axis. Hence write down the values of x for which:

(*a*) $x^2 - 3x + 2 > 0$ (*b*) $x^2 - 3x + 2 < 0$

2 Draw the graph of $y = x^2 + 2x - 3$. Write down the values of x where the graph lies above the x-axis. Hence write down the values of x for which:

(*a*) $x^2 + 2x - 3 \geq 0$ (*b*) $x^2 + 2x - 3 < 0$

3 By drawing the graph of $y = x^2 - 3x - 4$ find the values of x for which $x^2 - 3x - 4 > 0$.

4 By drawing the graph of $y = x^2 + 4x - 5$ find the values of x for which $x^2 + 4x - 5 \leq 0$.

135 Using quadratic inequalities to describe regions

You saw in Chapter 12 that $y > x + 2$ and $y < x + 2$ represented the regions on either side of the line $y = x + 2$. A similar situation arises whereby $y > x^2 - 5x + 4$ and $y < x^2 - 5x + 4$ represent the regions above and below the curve $y = x^2 - 5x + 4$.

■ *Example:* Draw the graph of $y = x^2 - 5x + 4$ and show the regions for which (*a*) $y > x^2 - 5x + 4$ and (*b*) $y < x^2 - 5x + 4$.

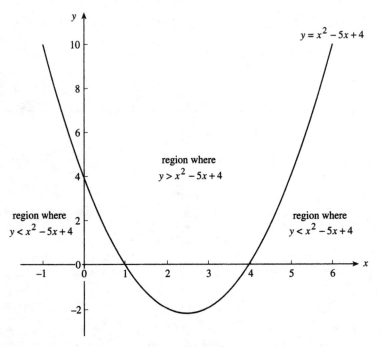

Fig. 59

Note: the point $(2, 3)$ lies in the region above the curve $y = x^2 - 5x + 4$ shown in Fig. 59 and when substituting the coordinates of this point you should find that the y-coordinate, 3, *is greater than* $2^2 - 5 \times 2 + 4 = -2$.

The point $(-1, 5)$ lies in the region below the curve and when substituting you should find that the y value, 5, *is less than* $(-1)^2 - 5 \times (-1) + 4 = 10$.

EXERCISE 49

1 Draw the graph of $y = x^2 - 4x + 3$ and show the regions for which:

(a) $y > x^2 - 4x + 3$ and (b) $y < x^2 - 4x + 3$

Check that the point $(1, 5)$ lies in the region above the graph.

2 Draw the graph of $y = x^2 - 4x - 5$ and show the regions for which:

(a) $y > x^2 - 4x - 5$ and (b) $y < x^2 - 4x - 5$

Check that the point $(1, 5)$ lies in the region above the graph.

3 Draw the graphs of $y = x^2$ and $y = 2x + 3$.
Shade the region where $y > x^2$ and $y < 2x + 3$.
Check that the point $(1, 3)$ lies in this region.

4 Draw the graphs of $y = x^2 - 3x$ and $y = 2x - 4$.
Shade the region where $y < x^2 - 3x$ and $y > 2x - 4$.
Find a point in the region where $y > x^2 - 3x$, $y < 2x - 4$ and $y > 0$.

17

Quadratic Equations

136 ALGEBRAICAL SOLUTION

Plotting the graph of a quadratic function led logically to the solution of a quadratic equation – i.e., an equation of the second degree. The solution by this method is useful and illuminating, but as a method of solving a quadratic equation it is cumbersome, and the accuracy obtainable is limited. An algebraical solution which is certain, universally applicable and capable of any required degree of accuracy is needed.

A quadratic equation was solved for the first time when, in Section 123, from the curve of $y = x^2$ it was found that if $x^2 = 3$, the corresponding values of x were $+1.73$ and -1.73. The reasoning may be stated thus:

$$x^2 = 3.$$
$$\therefore \qquad x = \pm\sqrt{3}$$
$$\text{and} \qquad x = +1.73 \text{ or } -1.73.$$

This is the simplest form of a quadratic equation. It involves the operation of finding a square root; hence the term 'root' as applied to the solution of an equation.

In Section 130 an important step forward was made. The points where the curve of $(x - 1)^2 - 4$ cuts the x-axis, and the function was therefore equal to zero, were found, and the corresponding values of x noted, namely 3 and -1.

This means that for these values of x, the expression

$$(x - 1)^2 - 4 = 0.$$

They are therefore the roots of this equation.
Let the equation be written in the form:

$$(x - 1)^2 = 4 \tag{A}$$

Then algebraically it is of the same form as the equation above, $x^2 = 3$. You can proceed with the algebraical solution on the same lines.

Taking the square roots of each side:

$$x - 1 = \pm 2.$$
$$\therefore \qquad x - 1 = +2 \text{ or } x - 1 = -2,$$
whence $\qquad x = 3 \text{ or } x = -1.$

The form marked (A) is the one to which, ultimately, all quadratic equations are reduced; the object always is to reach this form.

137 THE METHOD OF SOLUTION OF ANY QUADRATIC

It was shown that the expression $(x - 1)^2 - 4$ simplified to $x^2 - 2x - 3$.

$\therefore$ the equation which was solved could have been written in the form:

$$x^2 - 2x - 3 = 0 \tag{1}$$

If you start with this equation and wish to solve it, you need to get back to the form:

$$(x - 1)^2 - 4 = 0 \tag{2}$$

This is the converse operation of changing from (2) to (1), i.e. of obtaining the complete square (2) when you are given (1).

Two preliminary steps are necessary.

(1) Remove the constant to the right-hand side, as it does not help in finding the square.
(2) Divide throughout by the coefficient of x^2, if this is not unity.

After these steps you arrive at the form:

$$x^2 - 2x = 3.$$

It is now necessary to add to the left-hand side a number which produces a complete square. Remembering the work in Section 106, you get the following rule:

Add to each side the square of half the coefficient of x.

Then the above becomes

$$x^2 - 2x + (-1)^2 = 3 + 1.$$
$$\therefore \qquad (x - 1)^2 = 4,$$

and you proceed as before.

Here is the same method used to solve the equation:

$$x^2 - 6x + 5 = 0.$$
Then $$x^2 - 6x = -5.$$

Add to each side $\left(\dfrac{-6}{2}\right)^2$ – i.e., 3^2.

Then $$x^2 - 6x + (3)^2 = -5 + 9.$$
$$\therefore \qquad (x - 3)^2 = 4.$$

Thus you reach the desired form. Proceeding as before:

$$x - 3 = \pm 2.$$
$$\therefore \qquad x = 3 \pm 2.$$
Then $$x = 3 + 2 = 5$$
or $$x = 3 - 2 = 1.$$

$\therefore$ the solution is $x = 5$ or $x = 1$.

138 Solution of $2x^2 + 5x - 3 = 0$

Applying the preliminary steps (1) and (2) of the previous paragraph, you get in succession:

$$2x^2 + 5x = 3$$
and $$x^2 + \tfrac{5}{2}x = \tfrac{3}{2}.$$

Half the coefficient of x is $\tfrac{5}{4}$.
Adding $\left(\tfrac{5}{4}\right)^2$ you get:

$$x^2 + \tfrac{5}{2}x + \left(\tfrac{5}{4}\right)^2 = \tfrac{3}{2} + \left(\tfrac{5}{4}\right)^2.$$
$$\therefore \qquad \left(x + \tfrac{5}{4}\right)^2 = \tfrac{3}{2} + \tfrac{25}{16} = \tfrac{49}{16}.$$

Taking square roots of both sides:

$$x + \tfrac{5}{4} = \pm\tfrac{7}{4}.$$
$$\therefore \qquad x = -\tfrac{5}{4} \pm \tfrac{7}{4}$$
and $$x = -\tfrac{5}{4} + \tfrac{7}{4} = \tfrac{2}{4}$$
or $$x = -\tfrac{5}{4} - \tfrac{7}{4} = -\tfrac{12}{4}.$$
$$\therefore \qquad x = \tfrac{1}{2} \text{ or } x = -3.$$

139 WORKED EXAMPLES

■ *Example 1:* Solve the equation $x^2 - x - 1 = 0$.

Transposing

$$x^2 - x = 1.$$

Adding $(-\tfrac{1}{2})^2$

$$x^2 - x + (\tfrac{1}{2})^2 = 1 + \tfrac{1}{4}$$
or $$(x - \tfrac{1}{2})^2 = \tfrac{5}{4}.$$

$$\therefore \qquad x - \tfrac{1}{2} = \pm \frac{\sqrt{5}}{2}$$

and $$x = \tfrac{1}{2} \pm \frac{\sqrt{5}}{2}\,.$$

$$\therefore \qquad x = \tfrac{1}{2} + \frac{\sqrt{5}}{2} = \frac{1 + 2.236}{2} = 1.618$$

or $$x = \tfrac{1}{2} - \frac{\sqrt{5}}{2} = \frac{1 - 2.236}{2} = -0.618$$

$$\therefore \qquad x = 1.618 \text{ or } x = -0.618 \text{ (both approx.).}$$

■ *Example 2:* Solve the equation $3x^2 - 5x + 1 = 0$.

Applying preliminary steps, Section 137:

$$3x^2 - 5x = -1$$
$$x^2 - \tfrac{5}{3}x = -\tfrac{1}{3}.$$

Adding $(-\tfrac{5}{6})^2$:

$$x^2 - \tfrac{5}{3}x + (\tfrac{5}{6})^2 = -\tfrac{1}{3} + \tfrac{25}{36}.$$
$$\therefore \qquad (x - \tfrac{5}{6})^2 = \tfrac{13}{36}.$$

Taking square roots

$$x - \tfrac{5}{6} = \pm \frac{3.606}{6}$$

$$\therefore \qquad x = \frac{5 \pm 3.606}{6}\,.$$

$$\therefore \qquad x = \frac{5 + 3.606}{6} = \frac{8.606}{6} = 1.434$$

or $\qquad x = \dfrac{5 - 3.606}{6} = \dfrac{1.394}{6} = 0.232.$

∴ solution is $\qquad x = 1.434$ or $x = 0.232$ (approx.).

■ *Example 3:* Solve the equation:

$$\frac{1}{x - 1} - \frac{1}{x + 2} = \frac{1}{16} .$$

First clear the fractions by multiplying throughout by the least common denominator, namely $16(x - 1)(x + 2)$.

Then $\quad 16(x + 2) - 16(x - 1) = (x - 1)(x + 2).$
∴ $\qquad 16x + 32 - 16x + 16 = x^2 + x - 2.$

Adding and transposing:

$\qquad -x^2 - x = -48 - 2$
or $\quad x^2 + x = 50.$

Adding $(\tfrac{1}{2})^2$:

$x^2 + x + (\tfrac{1}{2})^2 = 50 + \tfrac{1}{4} = \tfrac{201}{4}.$

∴ $\quad (x + \tfrac{1}{2})^2 = \pm \dfrac{\sqrt{201}}{2} = \dfrac{\pm 14.177}{2} .$

∴ $\qquad x = -\tfrac{1}{2} \pm \dfrac{14.177}{2}$

and $\qquad x = -\tfrac{1}{2} + \dfrac{14.177}{2} = \dfrac{13.177}{2} = 6.588$ (approx.)

or $\qquad x = -\tfrac{1}{2} - \dfrac{14.177}{2} = -\dfrac{15.177}{2} = -7.588$ (approx.).

∴ the solution is

$$x = 6.588 \text{ or } x = -7.588.$$

EXERCISE 50

Solve the following equations:

1 $\quad 3x^2 = 12$ $\qquad\qquad$ **2** $\quad 4x^2 - 1 = 0$

3 $\quad \dfrac{x^2}{16} = 9$ $\qquad\qquad$ **4** $\quad (x + 1)^2 - 4 = 0$

5 $(x - 3)^2 - 25 = 0$ **6** $(x + 5)^2 = 36$

7 $(x + \frac{1}{4})^2 = 1$ **8** $\dfrac{(x - 9)^2}{15} = 4$

9 $x^2 - 10x + 16 = 0$ **10** $x^2 + x - 12 = 0$
11 $x^2 - 2x - 15 = 0$ **12** $x^2 + 3x - 28 = 0$
13 $x(x - 4) = 32$ **14** $2x^2 - 7x + 6 = 0$
15 $2x^2 - 3x - 5 = 0$ **16** $3x^2 = 7x + 9$

17 $3x^2 + 1 = 5x$ **18** $\dfrac{6}{x} + \dfrac{3}{x + 2} = 2$

19 $3x - \dfrac{5}{x} = 14$ **20** $x - 2 = \dfrac{2}{x}$

21 $\dfrac{x - 9}{3} = \dfrac{x + 5}{x}$ **22** $\dfrac{1}{x - 1} - \dfrac{1}{x + 2} = \dfrac{1}{16}$

140 SOLUTION OF QUADRATIC EQUATIONS BY FACTORISATION

There is another method of solving quadratics; it is as follows.

First note that if n is any finite number

$$n \times 0 = 0,$$

i.e., the product of any finite number and zero is always zero.

Conversely, if the product of two factors is zero, then either of the factors may be zero.

For example, if $a \times b = 0$ then this is true if either $a = 0$, or $b = 0$.

Similarly if

$$(x - 1)(x - 3) = 0 \tag{A}$$

it follows from the above that this equation is satisfied if either $x - 1 = 0$ or $x - 3 = 0$.

But if $x - 1 = 0$, then $x = 1$, and if $x - 3 = 0$, then $x = 3$.

∴ equation (A) must be satisfied if either $x = 1$ or $x = 3$.

∴ 1, and 3, are the roots of the equation:

$$(x - 1)(x - 3) = 0$$

i.e., $\qquad\qquad x^2 - 4x + 3 = 0.$

If you wish to solve the equation $x^2 - 4x + 3 = 0$, you can do it by reversing the above steps; consequently you find the factors of $x^2 - 4x + 3$ and get:

$$(x - 1)(x - 3) = 0.$$

If it is possible to obtain factors, this method is easy. In equations which arise out of practical work this can seldom be done.

It is a valuable method in other ways, and you can use it in equations of higher degree. If, for example, you know that

$$(x - 1)(x - 3)(x - 4) = 0.$$

Then, by the above reasoning, this equation is satisfied by:

$$x - 1 = 0, \text{ whence } x = 1$$
$$x - 3 = 0, \text{ whence } x = 3$$
$$x - 4 = 0, \text{ whence } x = 4.$$

The product of the three factors is an expression of the third degree, since the term of highest degree will be x^3. The equation is therefore of the third degree, or a cubic equation.

141 Worked examples

■ *Example 1:* Solve the equation $x^2 - 2x = 15$.

Transposing
$$x^2 - 2x - 15 = 0.$$

Factorising
$$(x - 5)(x + 3) = 0.$$
$$\therefore \qquad\qquad x - 5 = 0 \text{ and } x = 5$$
$$\text{or} \qquad\qquad x + 3 = 0 \text{ and } x = -3.$$

∴ the solution is $x = 5$ or $x = -3$.

Note: Remember that the justification for this method requires that the right-hand side is zero. Thus you cannot use it in a case such as

$$(x - 6)(x - 2) = 4.$$

■ *Example 2:* Solve the equation $9x(x + 1) = 4$.

Simplifying and transposing
$$9x^2 + 9x - 4 = 0.$$

Factorising
$$(3x + 4)(3x - 1) = 0.$$
$$\therefore \qquad\qquad 3x + 4 = 0 \text{ and } x = -\tfrac{4}{3}$$
$$\text{or} \qquad\qquad 3x - 1 = 0 \text{ and } x = \tfrac{1}{3}.$$

∴ the solution is $x = -\tfrac{4}{3}$ or $x = \tfrac{1}{3}$.

EXERCISE 51

Solve the following equations by the method of factors:

1 $x(x - 3) = 0$

2 $x(x + 5) = 0$

3 $(x - 2)^2 = 0$

4 $(x - 1)(x - 2) = 0$

5 $(x + 4)(x - 1) = 0$

6 $2(x - 3)(x + 7) = 0$

7 $4(3x - 7)(2x + 11) = 0$

8 $x^2 - 9x + 20 = 0$

9 $x^2 + x = 6$

10 $x^2 + 2x = 35$

11 $x(x + 13) + 30 = 0$

12 $x(x - 4) = x + 66$

13 $(x - 8)(x + 4) = 13$

14 $2x^2 - 11x + 12 = 0$

15 $2x^2 - 3x - 5 = 0$

16 $3x^2 - 4x + 1 = 0$

17 $8x^2 - 14x - 15 = 0$

18 $24x^2 + 10x - 4 = 0$

19 $(x - 2)(x - 4)(x - 5) = 0$

20 $(2x - 1)(x + 3)(x + 2) = 0$

21 $(x - 1)(x^2 - 2x - 8) = 0$

22 $(2x - 5)(x^2 - 5x - 50) = 0$

23 $(x - a)(x - b) = 0$

24 $(2x - c)(x + d) = 0$

142 GENERAL FORMULA FOR THE SOLUTION OF A QUADRATIC EQUATION

You have seen that by simplification and transposing every quadratic can be written in a form such as:

$$2x^2 + 7x - 4 = 0,$$

in which there are three constants – namely:

(1) the coefficient of x^2,
(2) the coefficient of x,
(3) the term independent of x.

Consequently if you want to write down a general form for any quadratic, letters such as a, b, c can be chosen to represent these constants, so that you could write the general form as:

$$ax^2 + bx + c = 0.$$

The equation above is a special case of this in which $a = 2$, $b = 7$, $c = -4$.

If you solved the general quadratic, $ax^2 + bx + c = 0$, the roots would be in terms of a, b, c. You should then have a formula such that, by substituting the values of a, b and c in any special case, you should be able to write down the roots, and no actual solving would be necessary.

143 SOLUTION OF THE QUADRATIC EQUATION $ax^2 + bx + c = 0$

The method used is 'completing the square', as in Section 137:

$$ax^2 + bx + c = 0.$$

Using preliminary step (1) $ax^2 + bx = -c.$

Using preliminary step (2) $x^2 + \dfrac{b}{a}x = -\dfrac{c}{a}.$

Adding 'the square of half the coefficient of x'

$$x^2 + \frac{b}{a}x + \left(\frac{b}{2a}\right)^2 = \frac{b^2}{4a^2} - \frac{c}{a}.$$

$\therefore \qquad \left(x + \dfrac{b}{2a}\right)^2 = \dfrac{b^2 - 4ac}{4a^2}$ (adding the fractions).

Taking square roots

$$x + \frac{b}{2a} = \pm \frac{\sqrt{b^2 - 4ac}}{2a}.$$

$\therefore \qquad x = -\dfrac{b}{2a} \pm \dfrac{\sqrt{b^2 - 4ac}}{2a}$

or $\qquad x = \dfrac{-b \pm \sqrt{b^2 - 4ac}}{2a}$

or in full $\qquad x = \dfrac{-b + \sqrt{b^2 - 4ac}}{2a}$

or $\qquad x = \dfrac{-b - \sqrt{b^2 - 4ac}}{2a}.$

This formula should be carefully learnt: by using it, there is no real necessity for working out any quadratic equation. Before applying it, however, the equation to be solved must be written down with all the terms on the left side.

From this formula, x must have two values and every quadratic must have two distinct roots. (The case when $b^2 - 4ac = 0$ is an exception.) This fact was obvious in the solution by means of a graph.

144 Worked Examples

■ *Example 1:* Solve the equation $5x^2 + 9x = 2$.

Writing the equation in general form you get:

$$5x^2 + 9x - 2 = 0.$$

Using the formula

$$x = \frac{-b \pm \sqrt{b^2 - 4ac}}{2a}$$

you have for this example

$$a = 5, b = 9, c = -2.$$

Substituting

$$x = \frac{-9 \pm \sqrt{9^2 - 4 \times 5 \times -2}}{2 \times 5}$$

$$= \frac{-9 \pm \sqrt{81 + 40}}{10}$$

$$= \frac{-9 \pm \sqrt{121}}{10}$$

$$= \frac{-9 \pm 11}{10}$$

$$\therefore \quad x = \frac{-9 + 11}{10} = \frac{2}{10}$$

or

$$x = \frac{-9 - 11}{10} = \frac{-20}{10}$$

$$\therefore \quad x = 0.2 \text{ or } -2.$$

■ *Example 2:* Solve the equation $\dfrac{1}{x - 1} + \tfrac{2}{3} = \dfrac{2}{x - 3}$.

Clearing the fractions by multiplying throughout by $3(x - 1)(x - 3)$

$$3(x - 3) + 2(x - 1)(x - 3) = 6(x - 1)$$
$$3x - 9 + 2(x^2 - 4x + 3) = 6x - 6$$
$$3x - 9 + 2x^2 - 8x + 6 = 6x - 6.$$

Collecting like terms

$$2x^2 - 11x + 3 = 0.$$

Using the formula

$$x = \frac{-b \pm \sqrt{b^2 - 4ac}}{2a}$$

and putting $a = 2, b = -11, c = 3,$

you get $x = \dfrac{11 \pm \sqrt{121 - 24}}{4}$

$$= \frac{11 \pm \sqrt{97}}{4} = \frac{11 \pm 9.849 \text{ (app.)}}{4}.$$

$\therefore \qquad x = \dfrac{11 + 9.849}{4} = \dfrac{20.849}{4} = 5.212$

or $\qquad x = \dfrac{11 - 9.849}{4} = \dfrac{1.151}{4} = 0.288$

$\therefore$ the solution is $x = 5.212$ or $x = 0.288$.

EXERCISE 52

Solve the following equations by using the formula of Section 143:

1 $x^2 + 3x - 1 = 0$ **2** $x^2 - 5x + 2 = 0$

3 $x^2 - 0.4x = 1.6$ **4** $3x^2 - 5x + 1 = 0$

5 $2x^2 - 5x = 2$ **6** $5x^2 = 7x + 3$

7 $2x^2 + 12x - 7 = 0$ **8** $3x^2 = 8x + 8$

9 $4x^2 + 10x = 5$ **10** $4x(x + 2) = 9$

11 $4x^2 = (x + 4)(2 - x)$ **12** $0.9(x + 1) = 0.8 - x^2$

13 $3x^2 = 7x + 2$ **14** $\dfrac{2}{3x + 1} + \dfrac{1}{2x + 1} = \frac{2}{7}$

15 $0.5 + \dfrac{2}{x + 3} - \dfrac{11}{x + 7} = 0$ **16** $\dfrac{3}{x + 2} - \dfrac{1}{x - 3} = \dfrac{4}{x}$

17 $(x + 3)(x - 5) = 2x - 2$

145 PROBLEMS LEADING TO QUADRATICS

The following examples illustrate the method of solving such problems.

■ *Example 1:* The distance (h) metres which a body reaches in time t seconds when it is projected vertically upwards with velocity u m/s is given by the formula

$$h = ut - \tfrac{1}{2}gt^2.$$

If $u = 160$ and $g = 10$, find the time a body takes to rise 240 metres.

Substituting the given values in the formula

$$240 = 160t - 5t^2$$
$$5t^2 - 160t + 240 = 0$$
$$t^2 - 32t + 48 = 0$$
$$\therefore \qquad t = \frac{32 \pm \sqrt{32^2 - 4 \times 48}}{2}$$

$$= 16 \pm 4\sqrt{13}$$
$$= 16 \pm 4 \times 3.6 \text{ approx.}$$
$$t = 1.58 \text{ secs or } 30.4 \text{ secs.}$$

The two roots require consideration. Since every quadratic has two roots, when these give the answer to a problem you must examine the applicability of the roots to the problem. Sometimes it will be clear that both are not applicable, especially if one is negative. In this problem, when a body is projected vertically upwards its velocity decreases until it reaches its highest point, when it is zero. It then falls vertically and retraces its path. Therefore it will be at a given height twice, once when ascending and again when descending. The value $t = 1.84$ gives the time to reach 277.5 metres when going up, and it is at the same height when descending 30.2 seconds after starting.

■ *Example 2:* You travel a distance of 84 km by car. You find that, if on the return journey you increase your average speed by 4 km/h, you will take half an hour less. What was your average speed for the first part of the journey and how long did you take for the double journey?

Let x miles per hour be your average speed for the first journey.

Then the time for the first journey is $\dfrac{84}{x}$ hours.

The speed for the return journey is $(x + 4)$ miles per hour.

$\therefore$ time for the second journey is $\dfrac{84}{x + 4}$ hours.

But this is $\frac{1}{2}$ hour less than the first time.

$$\therefore \qquad \frac{84}{x} - \frac{84}{x + 4} = \tfrac{1}{2}.$$

Clearing fractions by multiplying by the least common denominator, namely $2x(x + 4)$.

$$168(x + 4) - 168x = x(x + 4)$$
$$168x + 672 - 168x = x^2 + 4x.$$

Collecting and transposing

$$x^2 + 4x - 672 = 0.$$

Using the formula

$$x = \frac{-b \pm \sqrt{b^2 - 4ac}}{2a}$$

or substituting $x = \dfrac{-4 \pm \sqrt{16 - (4 \times -672)}}{2}$

$$= \frac{-4 \pm \sqrt{2704}}{2}$$

$$= \frac{-4 \pm 52}{2}$$

$\therefore$ $x = \dfrac{-4 + 52}{2} = 24$

or $x = \dfrac{-4 - 52}{2} = -28.$

The negative root, although it satisfies the equation above, has no meaning for this problem.

$\therefore$ average speed for first journey is 24 km/h.

Time for first journey $= \frac{84}{24} = 3\frac{1}{2}$ h.
Time for second journey $= \frac{84}{28} = 3$ h.
$\therefore$ total time $= 6\frac{1}{2}$ hours.

EXERCISE 53

1   The sum of a number and its reciprocal is 2.9. Find the number.
2   The area of a rectangle is 135 mm^2 and its perimeter is 48 mm. What are the lengths of its sides?
3   Solve for $\dfrac{1}{R}$ the equation $\dfrac{1}{R^2} - \dfrac{5}{R} = 10.$

4 The relation between the joint resistance R and two resistances r_1 and r_2 in parallel is given by the formula $\dfrac{1}{R} = \dfrac{1}{r_1} + \dfrac{1}{r_2}$.
If R is 12 ohms and r_2 6 ohms greater than r_1, find r_1 and r_2.

5 A formula for finding the strength of a concrete beam is $bn^2 + 2am(n - c) = 0$. Solve this for n when $b = 4$, $a = 2$, $c = 8$, $m = 12.5$.

6 The formula giving the sag (D) in a cable of length L and span S is expressed by $L = \dfrac{8D^2}{3S} + S$. Find S when $L = 80$, $D = 2.5$.

7 The product of a number n and $2n - 5$ is equal to 250. What is the value of n?

8 There is an algebraical formula $s = \dfrac{n}{2}\{2a + (n - 1)d\}$. If $s = 140$, $a = 7$ and $d = 3$, find n.

9 Pythagoras's theorem states that the square on the diagonal of a rectangle is equal to the sum of the squares on the two sides. The diagonal of a particular rectangle is 25 mm long, and one side of the rectangle is 5 mm longer than the other. Find the sides of the rectangle.

10 The cost of a square carpet is £248 and the cost of another square carpet whose side is 3 m longer than that of the first is £387.50. If the cost per m^2 is the same for both carpets, find the area of each.

11 There is a number such that when it is increased by 3 and the sum is squared, the result is equal to 12 times the number increased by 16. What is the number?

12 Two adjacent sides of a rectangle are represented by $(x + 4)$ and $(x + 6)$. The area of the rectangle is equal to twice the area of the square whose side is x. What is the value of x?

13 Find the area of a rectangular plot of ground whose perimeter is 42 m and whose diagonal is 15 m.

14 The cost of boring a well is given by the formula, $C = 20x + \dfrac{1}{5}x^2$, where C is the cost in pounds and x is the depth in metres. If a well cost £4,000 to bore, how deep was it?

15 One number exceeds another by 4. The sum of their squares is 208. What are the numbers?

16 The formula for the sum of the first n whole numbers is $\frac{1}{2}n(n + 1)$. If the sum is 78, how many numbers are there?

146 SIMULTANEOUS EQUATIONS OF THE SECOND DEGREE

These are equations which involve two unknowns and which include terms of the second degree.

They are also called simultaneous quadratic equations.

The degree of a term is shown either by its index, or, if it contains two or more letters, by the sum of their indices.

x^2, xy, y^2 are terms of the second degree involving two unknowns, and any of these, together with terms of the first degree, may occur in the equations to be solved. Numerical coefficients do not affect the 'degree' of a term.

It is seldom possible to solve simultaneous quadratics unless they conform to certain types. In this book two of these types are considered.

147 FIRST TYPE. WHEN ONE OF THE EQUATIONS IS OF THE FIRST DEGREE

Example

$$x + y = 1 \tag{1}$$
$$3x^2 - xy + y^2 = 37 \tag{2}$$

You can always solve simultaneous quadratic equations if you can re-arrange one of the equations to find one letter in terms of the other, provided the resulting expression is linear.

For example, in the equations above, from (1) you get:

$$y = 1 - x.$$

You can now substitute this into equation (2).
Thus you get:

$$3x^2 - x(1 - x) + (1 - x)^2 = 37.$$

In this way you reach a quadratic, with one unknown, and you can solve this by the methods previously given.

Simplifying

$$3x^2 - x + x^2 + 1 - 2x + x^2 = 37.$$
$$\therefore \qquad 5x^2 - 3x - 36 = 0.$$

You can solve this by factorisation, or by using the formula:

$$x = \frac{-b \pm \sqrt{b^2 - 4ac}}{2a}.$$

On substituting $x = \dfrac{+3 \pm \sqrt{9 + 720}}{10}$

$$= \dfrac{3 \pm 27}{10} \ .$$

$\therefore \qquad x = \dfrac{30}{10} \text{ or } \dfrac{-24}{10}$

or $\qquad x = 3 \text{ or } -2.4.$

To find y substitute in

$$y = 1 - x.$$

(1) When $x = 3$,

$$y = 1 - 3 = -2.$$

(2) When $x = -2.4$,

$$y = 1 - (-2.4) = 3.4.$$

$\therefore$ the solutions are:

(1) $x = 3, y = -2.$
(2) $x = -2.4, y = 3.4.$

You must arrange the solutions in corresponding pairs.

EXERCISE 54

Solve the following equations:

1 $\quad x - y = 2$
$\quad x^2 + xy = 60$

2 $\quad x + y = 7$
$\quad 3x^2 + xy - y^2 = 81$

3 $\quad 2x + y = 5$
$\quad 5x^2 - 3xy = 14$

4 $\quad 3x + y = 8$
$\quad x^2 - 5xy + 8y = 36$

5 $\quad x + y + 1 = 0$
$\quad 3x^2 - 5y^2 - 7 = 0$

6 $\quad 2x + 3y = 14$
$\quad 4x^2 + 2xy + 3y^2 = 60$

7 $\quad 2x^2 - 5x + 4xy = 60$

$\quad 3x - y = 9$

8 $\quad 2x + 3y = 5$

$\quad \dfrac{5}{y + 3} - \dfrac{9}{x} = 1$

9 $\quad 3x + y = 25$

$\quad xy = 28$

10 $\quad 2x - y = 8$

$\quad \dfrac{x^2}{9} + \dfrac{y^2}{16} = 5$

11 $\quad \dfrac{1}{x} + \dfrac{1}{y} = \tfrac{3}{4}$

$\quad 3x - y = 2$

148 SECOND TYPE. SYMMETRIC EQUATIONS

■ *Example 1:* An example of this type is

$$x + y = 19 \tag{1}$$
$$xy = 84 \tag{2}$$

These two equations are called 'symmetric' because, if the letters x and y are interchanged throughout, the equations are unaltered. Other equations which are not strictly symmetric can be solved by the same method as these, and so are included.

In these particular equations number (1) is of the first degree, and so the previous method can be used; but this is an easy example to illustrate the special solution which can be employed with this type.

The aim of the method is to find the value of $x - y$. Then, since the value of $x + y$ is known, the rest of the solution is easy.

We are given

$$x + y = 19 \tag{1}$$
$$xy = 84 \tag{2}$$

Squaring both sides of (1)

$$x^2 + 2xy + y^2 = 361 \tag{3}$$

From (2)

$$4xy = 336 \tag{4}$$

Subtracting equation (4) from equation (3) you get

$$x^2 - 2xy + y^2 = 25$$

$\therefore \qquad\qquad x - y = \pm 5.$

From (1) $\qquad\qquad x + y = 19.$

Adding $\qquad\qquad 2x = 19 \pm 5 = 24$ or $14.$

$\therefore \qquad\qquad x = 12$ or $7.$

Subtracting $\qquad\qquad 2y = 19 - (\pm 5)$
$\qquad\qquad\qquad\qquad = 14$ or $24.$

$\therefore \qquad\qquad y = 7$ or $12.$

Or using $\qquad\qquad xy = 84$

(1) when $\qquad\qquad x = 12, y = 7,$

(2) when $\qquad\qquad x = 7, y = 12.$

The symmetry of the two equations appears in the solution.

■ *Example 2:* Solve the equations:

$$x^2 + y^2 = 89 \qquad (1)$$
$$xy = 40 \qquad (2)$$

Both equations are of the second degree. The aim in this case is to obtain both $x + y$ and $x - y$.

From (2)

$$2xy = 80 \qquad (3)$$

Adding equations (1) and (3)

$$x^2 + 2xy + y^2 = 169$$
∴ $\qquad x + y = \pm 13.$

Subtracting equation (3) from equation (1)

$$x^2 - 2xy + y^2 = 9.$$
∴ $\qquad x - y = \pm 3.$

Putting the two pairs of equations together:

$$x + y = \pm 13.$$
$$x - y = \pm 3.$$

Adding $\qquad 2x = \pm 13 \pm 3 = \pm 16 \text{ or } \pm 10.$
∴ $\qquad x = \pm 8 \text{ or } \pm 5.$

To get corresponding values of x you now substitute in the equation

$$xy = 40.$$

∴ when $\qquad x = +8, -8, +5, -5$
$$y = +5, -5, +8, -8.$$

Putting these solutions together, the final solution is:

$$x = +8, y = +5$$
$$x = -8, y = -5$$
$$x = +5, y = +8$$
$$x = -5, y = -8.$$

Note: In the above examples the sum and difference of two numbers, $x + y$, $x - y$, are used. The same method would be used if you were given an example such as $x + 3y$. You should then aim to find $(x + 3y)^2$, $(x - 3y)^2$, and ultimately $x - 3y$. There are many variations in this type of solution. In some, such as

$$x^2 + y^2 = a \qquad (1)$$
$$x + y = b \qquad (2)$$

xy is not given, but you can find it by squaring equation (2) and subtracting equation (1).

EXERCISE 55

Solve the following equations:

1 $x + y = 8$
$xy = 15$

2 $x + y = 9$
$3xy - 42 = 0$

3 $x - y = 5$
$xy = 24$

4 $x^2 + y^2 = 17$
$x + y = 5$

5 $x - 2y = 2$
$xy = 12$

6 $x^2 + y^2 = 17$
$xy = 4$

7 $x^2 - xy + y^2 = 67$
$xy = 18$

8 $x^2 - xy + y^2 = 43$
$x + y = 8$

9 $x^2 + y^2 = 45$
$x^2 - xy + y^2 = 27$

10 $2x + 3y = 9$
$xy = 3$

149 SOLVING QUADRATIC INEQUALITIES

As you saw in Chapter 16, plotting a graph of the associated quadratic function illustrated one way in which the solution of a quadratic inequality could be obtained. However, as with the solution of quadratic equations, this method can be rather cumbersome unless you have a graphics calculator. Here is another method, using factorisation.

■ *Example 1:* Solve the quadratic inequality $x^2 - 6x + 8 < 0$.

The left hand side of the inequality can be factorised as $(x - 2)(x - 4)$. So you now want to find the values of x for which $(x - 2)(x - 4) < 0$.

You want the product of the two brackets to be a negative number i.e. < 0. This can only happen if one of the brackets is a positive number and the other is a negative number. So you want $(x - 2) > 0$ and $(x - 4) < 0$ or $(x - 2) < 0$ and $(x - 4) > 0$.

The best way to handle these pairs of simultaneous inequalities is to use the number line.

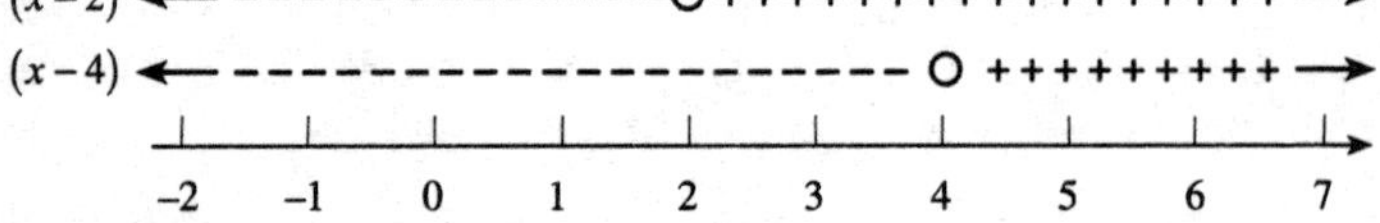

Fig. 60

In Fig. 60, the fact that $(x - 2)$ is positive when $x > 2$ and negative when $x < 2$ is shown by using the $+$ and $-$ signs. In the same way $(x - 4)$ is positive when $x > 4$ and negative when $x < 4$. For $(x - 2)(x - 4)$ to be less than 0 you need the values of x where you have one positive and one negative bracket. This only happens for values of x between 2 and 4, so the solution of $x^2 - 6x + 8 < 0$ is where $2 < x < 4$.

If you had wanted to find where $x^2 - 6x + 8 > 0$ you would have been looking for the values of x where the product of $(x - 2)$ and $(x - 4)$ was positive. This occurs *either* when both brackets are positive or when both brackets are negative. This happens when $x > 4$ or when $x < 2$. So the solution of $x^2 - 6x + 8 > 0$ is where $x < 2$ or where $x > 4$.

The advantage of this method is that you can extend it to cater for more complex situations where several brackets are involved. The next example shows how a cubic inequality can be solved.

■ *Example 2:* Solve the inequality $(2x - 1)(x + 2)(3 - x) > 0$.

The product of the three brackets has to be a positive number. This occurs *either* if all three of the brackets are positive *or* if one bracket is a positive number and the other two are negative numbers.

So on the number line diagram we want three $+$'s or one $+$ and two $-$'s.

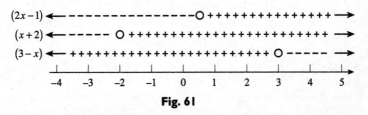

Fig. 61

All three brackets are positive for values of x which are between 0.5 and 3. You have one positive and two negative brackets for values of x less than -2. So the solution of the inequality is where $0.5 < x < 3$ or where $x < -2$.

If the inequality has been $(2x - 1)(x + 2)(3 - x) \geq 0$ it would have been necessary to include the values -2, 0.5 and 3, i.e. $x \leq -2$ or $0.5 \leq x \leq 3$.

EXERCISE 56

1 Use the number line method to solve the inequality:

(a) $(x - 2)(x - 3) > 0$ (b) $(x + 2)(x - 4) < 0$
(c) $(x + 2)(x + 3) \geq 0$

2 Factorise the quadratic expression and then use the number line method to solve the inequality:

(a) $x^2 - 3x + 2 > 0$ (b) $x^2 - 6x + 5 > 0$
(c) $x^2 - 3x - 4 \geq 0$

3 Use the number line method to solve the inequality:

(a) $(2x - 2)(3 - x) > 0$ (b) $(4x - 1)(x + 2) < 0$
(c) $(2x + 1)(3x - 1) \geq 0$

4 Use the number line method to solve the inequality:

(a) $(x - 2)(x - 3) > 0$ (b) $(2x + 1)(4x - 1)(x - 3) < 0$

5 Solve the inequality:

(a) $x^2 - 3x < 0$ (b) $5x - x^2 \geq 0$
(c) $(x - 2)(x - 2)(x - 5) > 0$

18

Indices

150 THE MEANING OF AN INDEX

In Section 20 it was shown that the product of a number of equal factors such as $a \times a \times a \times a$ can be written in the form a^4, in which 4 is an index which indicates the number of factors.

Generalising this, if there are n such equal factors, where n is a positive integer, then a^n will be defined as follows:

$$a^n = a \times a \times \ldots \text{ to } n \text{ factors}$$
and a^n is called the nth power of a.

151 LAWS OF INDICES

It was also shown in Sections 21, 22, 23 that when operations such as multiplication and division of powers of a number are performed, the laws which govern these operations can be deduced from the definitions of a power and an index given in Section 150.

General proofs of these laws will now be given.

I First law of indices: the law of multiplication

The special cases given in Section 21 lead to the general law as follows:

Let m and n be any positive integers.
By definition

$$a^m = a \times a \times a \times \ldots \text{ to } m \text{ factors}$$
and $a^n = a \times a \times a \times \ldots \text{ to } n \text{ factors}.$

Then $a^m \times a^n = (a \times a \times a \times \ldots$ to m factors$) \times (a \times a \times a \times \ldots$ to n factors$)$.

But when two groups of factors are multiplied the factors in the groups are associated as one group of factors to give the product (see Section 21).

$$\therefore \quad a^m \times a^n = a \times a \times a \times a \times \ldots \text{ to } (m + n) \text{ factors}$$
$$= a^{m+n} \text{ (by definition).}$$

$\therefore$ the first law of indices is

$$a^m \times a^n = a^{m+n}$$

The law is clearly true when the product involves more than two powers. Thus $a^m \times a^n \times a^p = a^{m+n+p}$.

II   Second law of indices: the law of division

To find the value of $a^m \div a^n$.

With the same definition as before, and proceeding as in the special cases of Section 23.

$$a^m \div a^n = \frac{a \times a \times a \times \ldots \text{ to } m \text{ factors}}{a \times a \times a \times \ldots \text{ to } n \text{ factors}}.$$

After cancelling the n factors of the denominator with n corresponding factors in the numerator, there are left in the numerator $m - n$ factors.

$$\therefore a^m \div a^n = a^{m-n}$$

Note: This proof assumes that m is greater than n. If m is less than n, there are $n - m$ factors left in the denominator.

$$\therefore \text{ if } n > m \qquad\qquad a^m \div a^n = \frac{1}{a^{n-m}}.$$

This case will be examined later.

III   Third law of indices: the law of powers

By the definition of Section 150 such an expression as $(a^4)^3$ means the third power of a^4, or

$$(a^4)^3 = a^4 \times a^4 \times a^4$$
$$= a^{4+4+4} \text{ (by first index law)}$$
$$= a^{4 \times 3}$$
$$= a^{12}.$$

In general, if m and n are any positive integers, by definition

$$(a^m)^n = a^m \times a^m \times a^m \quad \ldots \text{to } n \text{ factors}$$
$$= a^{m+m+m} \quad \ldots \text{to } n \text{ terms (first index law)}$$
$$= a^{m \times n}.$$

∴ the law of powers is

$$(a^m)^n = a^{m \times n}.$$

For powers of a product such as $(ab)^n$, see Section 22. In a similar way it may be shown that

$$(ab)^n = a^n \times b^n.$$

EXERCISE 57

REVISION EXERCISE IN INDICES

Write down the values of the following:

1 $2a^4 \times 3a^5$ **2** $a^6 \times a^6$

3 $\frac{1}{2}a^5 \times \frac{1}{4}a^3$ **4** $2 \times 2^2 \times 2^3 \times 2^4$

5 $4 \times 4^2 \times 4^3$ **6** $3a^3b^4 \times 2a^5b^3$

7 $x^4y^5z \times x^2y^7z^3$ **8** $x^{m+1} \times x^{m-1}$

9 $a^{m+n} \times a^{m-n}$ **10** $a^pb^q \times a^{p+q} \times b^{p-q}$

11 $a^{n+3} \times a^{n-3}$ **12** $a^m \times a^n \times a^4$

13 $a^8 \div a^5$ **14** $5x^6 \div 10x^3$

15 $36a^{10} \div 12a^5$ **16** $2^{10} \div 2^4$

17 $3^5 \div 3^2$ **18** $(-x^7) \div (-x^2)$

19 $15a^4b^2 \div -3ab$ **20** $a^{2p} \div a^p$

21 $12x^{2p}y^{2q} \div 3x^py^q$ **22** $\dfrac{a^{10} \times a^4}{a^7}$

23 $\dfrac{x^8}{x^4} \times \dfrac{x^3}{x^6}$ **24** $a^{2n} \div a^{n-1}$

25 $a^{m+n} \div a^{m-n}$ **26** $a^{n+4} \div a^{n-2}$

27 $\dfrac{4a^3}{3b^2} \div \dfrac{6a^2}{4b^3}$ **28** $\dfrac{a}{b} \div \dfrac{b}{a}$

29 $(2^3)^2$ **30** $(3^2)^3$

31 $(x^5)^2$ **32** $(x^2)^5$

33 $(a^4)^4$ **34** $(2a^2b^3)^2$

35 $\left(\dfrac{x^4y^5}{2y^2}\right)^3$ **36** $(a^p)^3$

37 $(x^4)^n$ **38** $(3a^{2p})^3$

39 $\sqrt{a^4}$ **40** $\sqrt{x^8}$

41 $\sqrt{x^{16}}$ **42** $\sqrt{9a^6}$

43 $\sqrt{a^4b^2}$ **44** $\sqrt{\dfrac{x^6}{y^4}}$

45 $\sqrt{\dfrac{9a^8}{4b^6}}$ **46** $\sqrt{x^{2n}}$

152 EXTENSION OF THE MEANING OF AN INDEX

It has so far been assumed that all indices are positive integers. The definition of a^n, namely:

$$a^n = a \times a \times a \times \quad \ldots \text{to } n \text{ factors}$$

is meaningless unless you assume that n is a positive integer.

But algebra generalises, and you must therefore consider the possibility of attaching a meaning to an index in all cases.

153 GRAPH OF 2^x

As a first step choose a suitable positive number, say 2, and plot some of the powers of it – i.e., draw the graph of 2^x, in which x represents any index. Calculating the values of these powers for some of the smaller integral values of x, you get a table of values as follows:

x	1	2	3	4
$y = 2^x$	2	4	8	16

When these points are plotted they appear to be on a smooth curve, as drawn in Fig. 62.

If you are justified in assuming that this curve is continuous and such that all the points on it satisfy the equation $y = 2^x$, then it follows that if you take any point on the curve between the plotted points, its coordinates must also satisfy the law.

Therefore if any point, A, is taken on the curve at which $x = 1.5$ and $y = 2.8$, then it follows that $2^{1.5} = 2.8$.

Again at B, where $x = 3.2$ and $y = 9$, $2^{3.2} = 9$.

Hence, if the assumption is correct, that the curve is a continuous one, and the coordinates of any point on it satisfy $y = 2^x$, then it may be concluded that any number within the limits plotted can be expressed as a power of 2, and conversely any number can be used as the index of some power of 2.

A similar curve could be drawn exhibiting the powers of any other number – say, 10 – i.e., you could draw the curve of

$$y = 10^x.$$

Thus you could express any number as a power of 10.

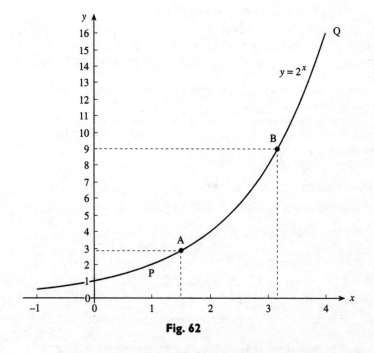

Fig. 62

Reverting to the curve PQ (Fig. 62), since it was plotted by taking values of x from $x = 1$, the curve began at P. If it is produced towards the y-axis in the way it seems to curve, it will apparently cut the y-axis at the point where $y = 1$.

This suggests that:

(1) The value of 2^x when $x = 0$ is 1, i.e., $2^0 = 1$.
(2) The portion of the curve thus drawn should show values of 2^x between $x = 0$ and $x = 1$.

For example, when $x = \frac{1}{2}$, 2^x is approximately 1.4, i.e., $2^{\frac{1}{2}} = 1.4$ approximately.

Again the curve evidently does not end at the y-axis, but can be further produced. This portion will correspond to negative values of x. It follows that values of 2^x can be found when the index is negative.

154 ALGEBRAICAL CONSIDERATION OF THE EXTENSION OF THE MEANING OF INDICES

From the graph of powers it may be inferred that powers of a number exist whether the index is integral or fractional, positive or negative. But how can such indices be interpreted algebraically?

In seeking to find meanings for the new forms of indices there is one fundamental guiding principle, namely:

Every index must obey the laws of indices as discovered for positive integral numbers otherwise it cannot be considered as an index.

155 FRACTIONAL INDICES

The simplest example of a fractional index is $a^{\frac{1}{2}}$.

To find a meaning for $a^{\frac{1}{2}}$.
Since $a^{\frac{1}{2}}$ must conform to the laws of indices
∴ by the first law

$$a^{\frac{1}{2}} \times a^{\frac{1}{2}} = a^{\frac{1}{2}+\frac{1}{2}}$$
$$= a^1 \text{ or } a.$$

∴ $a^{\frac{1}{2}}$ must be such a number that on being multiplied by itself the product is a.

But such a number is defined arithmetically as the square root of a.

∴ $a^{\frac{1}{2}}$ must be defined as $\sqrt{a}$.

As an example $2^{\frac{1}{2}} = \sqrt{2} = 1.4142$ approx.

This agrees with the value of 1.4 found from the graph (Fig. 62). The reasoning above clearly holds in all such cases, and so it follows that in general, if n is any positive integer

$$a^{\frac{1}{n}} = \sqrt[n]{a}.$$

To find a meaning for $a^{\frac{2}{3}}$.

Applying the first law of indices:

$$a^{\frac{2}{3}} \times a^{\frac{2}{3}} \times a^{\frac{2}{3}} = a^{\frac{2}{3}+\frac{2}{3}+\frac{2}{3}}$$
$$= a^2.$$
$$\therefore \qquad a^{\frac{2}{3}} = \sqrt[3]{a^2}.$$

Applying the reason generally, it follows that if m and n are any positive integers

$$a^{\frac{m}{n}} = \sqrt[n]{a^m}.$$

Thus
$$a^{\frac{3}{4}} = \sqrt[4]{a^3}.$$
$$a^{\frac{2}{5}} = \sqrt[5]{a^2}.$$

Indices which are in decimal form can be changed to vulgar fractions. Thus:

$$a^{0.25} = a^{\frac{1}{4}}$$
$$= \sqrt[4]{a}.$$

156 To find a meaning for a^0

Using the second law of indices, and n being any number,

$$a^n \div a^n = a^{n-n}$$
$$= a^0.$$
But
$$a^n \div a^n = 1.$$
$$\therefore \qquad a^0 = 1.$$

This confirms the conclusion reached in Section 153 from the graph. It should be noted that a represents any finite positive number. Therefore $a^0 = 1$, whatever the value of a.

Graphically, if curves of a^x are drawn for various values of a, as that of 2^x was drawn in Fig. 62, all these curves will pass through the point on the y-axis which is unit distance from 0.

157 Negative indices

In considering the curve of 2^x in Fig. 62, the conclusion was reached that the curve could be continued for negative values of x.

But what meaning can be given algebraically to a negative index?

Consider a^{-1}.

This must obey the laws of indices.

∴ by the first law

$$a^{-1} \times a^{+1} = a^{-1+1}$$
$$= a^0$$
$$= 1.$$

Dividing by a^{+1} you get

$$a^{-1} = \frac{1}{a},$$

i.e., a^{-1} must mean 'the reciprocal of a'.

With the same reasoning

$$a^{-2} = \frac{1}{a^2}$$

$$a^{-3} = \frac{1}{a^3},$$

and in general

$$a^{-n} \times a^{+n} = a^{-n+n}$$
$$= a^0$$
$$= 1.$$

∴ $$a^{-n} = \frac{1}{a^n}.$$

∴ a^{-n} is defined as the reciprocal of a^n.

You should note the following examples.

$$2a^{-3} = \frac{2}{a^3} \text{ (negative index applies to } a \text{ only)}$$

$$a^{-\frac{1}{2}} = \frac{1}{\sqrt{a}}$$

$$\frac{1}{a^{-1}} = a$$

and generally $$\frac{1}{a^{-n}} = a^n.$$

You should note as a working rule that when a power of a number is transferred from the numerator of a fraction to the denominator, or vice-versa, the sign of the index is changed.

If n is a positive number and $a > 1$, a^n increases as n increases.

$\therefore a^{-n}$ or $\dfrac{1}{a^n}$ decreases as n increases.

Consequently the curve of 2^x will approach closer to the x-axis for negative values of x.

This agrees with the course which the curve in Fig. 62 appeared to be taking.

158 STANDARD FORMS OF NUMBERS

Indices are useful in writing numbers in standard form. This helps you to express clearly and concisely numbers which are very large or very small.

You can see how to write them from the following examples:

$$26 = 2.6 \times 10^1$$
$$260 = 2.6 \times 10^2$$
$$2600 = 2.6 \times 10^3$$
$$26\,000 = 2.6 \times 10^4$$
$$260\,000 = 2.6 \times 10^5 \text{ and so on.}$$

When a number is written in standard form, the number is written as a decimal number between 1 and 10, (i.e., with one digit only retained in the whole number part) and this is multiplied by a power of 10 to make it equal to the original number.

It should be noted that the index of the power of 10 is equal to the number of times the decimal point has to be moved to convert the number in standard form back to the original number. If the point has to be moved to the right you obtain a positive index. If it is moved to the left, the index is negative.

Thus if 547 000 000 is to be written in standard form, 5 only is retained in the whole number, with the following numbers included after the decimal point, i.e., 5.47. To regain the original number the decimal point would have to be moved 8 times to the right, so we find that

$$547\,000\,000 = 5.47 \times 10^8.$$

Note that here, since the index is positive, there is no need to write down the trailing (insignificant) zeros of the decimal number in the standard form. These are automatically accounted for by the power of 10.

Very small numbers between 0 and 1 can similarly be changed into standard form but now the index of 10 must be negative.

Thus
$$0.26 = 2.6 \div 10 \text{ or } 2.6 \times 10^{-1}$$
$$0.026 = 2.6 \div 10^2 \text{ or } 2.6 \times 10^{-2}$$
$$0.0026 = 2.6 \div 10^3 \text{ or } 2.6 \times 10^{-3}$$
$$0.000\,26 = 2.6 \div 10^4 \text{ or } 2.6 \times 10^{-4}, \text{ etc.}$$

You should note that the numerical part of the index is one more than the number of zeros following the decimal point in the original number.

If the number is overall a negative one, then the standard form converts it to a number which is between -10 and -1. The overall sign is independent of the index, for example.

$$-135\,000 = -1.35 \times 10^5$$
$$-0.000\,257 = -2.57 \times 10^{-4}$$

159 OPERATIONS WITH STANDARD FORMS

In these operations you must observe the rules of indices.

■ *Examples:*

(1) $(1.2 \times 10^4) \times (2.3 \times 10^3) = (1.2 \times 2.3) \times (10^4 \times 10^3)$
$$= 2.76 \times 10^7.$$

(2) $(4.8 \times 10^8) \div (1.6 \times 10^{-3}) = (4.8 \div 1.6) \times (10^8 \div 10^{-3})$
$$= 3 \times 10^{11}.$$

EXERCISE 58

Note: When necessary in the following examples take $\sqrt{2} = 1.414$, $\sqrt{3} = 1.732$, $\sqrt{5} = 2.236$, $\sqrt{10} = 3.162$.

1 State what meanings can be given to the following:

(1) $3^{\frac{1}{4}}$ (2) $8^{\frac{1}{6}}$ (3) $a^{\frac{1}{3}}$
(4) $2^{0.5}$ (5) $3^{0.5}$ (6) $a^{0.75}$

2 State what meanings can be given to the following:

(1) $10^{\frac{3}{4}}$ (2) $4^{\frac{2}{3}}$ (3) $a^{\frac{5}{4}}$
(4) $10^{1.5}$ (5) $2^{2.5}$ (6) $10^{1.25}$

3 Write down as simply as possible the meaning of:

(1) $4^{\frac{3}{6}}$ (2) $8^{\frac{3}{4}}$ (3) $a^{1.75}$
(4) $a^{\frac{7}{3}}$ (5) $a^{1.2}$ (6) $a^{0.2}$

4 Find the numerical values of:

(1) $4^{1.5}$ (2) $16^{1.25}$ (3) $9^{1.5}$
(4) $100^{\frac{3}{2}}$ (5) $10^{\frac{3}{2}}$ (6) $(\frac{1}{4})^{0.5}$

5   Find the values of 3^{-2}, 3^{-1}, 3^{0}, 3^{1}, $3^{1.5}$, 3^{2}, $3^{2.5}$. Plot these and draw the curve which contains the points.

6   Find the values of:

(1) $2^2 \times 2^{\frac{1}{2}}$ (2) $3 \times 3^{\frac{1}{2}} \times 3^{\frac{3}{2}}$
(3) $10^{\frac{1}{2}} \div 10^{\frac{3}{2}}$ (4) $a^{\frac{2}{3}} \times a^{\frac{3}{4}}$
(5) $2^{\frac{5}{2}} \div 2^{\frac{3}{2}}$ (6) $a^{1.5} \div a^{0.3}$

7   Write down the meanings with positive indices of:

$$4^{-1}, 3a^{-2}, 2^{-\frac{1}{2}}, \frac{1}{3^{-1}}, \frac{3}{a^{-2}}, 10^{-3}.$$

8   Write with positive indices:

(1) x^{-3} (2) $x^{-\frac{1}{2}}$ (3) $\dfrac{1}{x^{-3}}$

(4) $(x^{-3})^2$ (5) $\left(\dfrac{1}{x}\right)^{-1}$ (6) $\dfrac{1}{2x^{-\frac{1}{2}}}$

9   Find the values of:

(1) $8^{\frac{2}{3}}$ (2) $25^{\frac{3}{2}}$ (3) $(10^2)^{\frac{3}{2}}$

(4) $(5^{-3})^2$ (5) $\dfrac{2}{2^{-3}}$ (6) $81^{\frac{1}{4}}$

10   Find the values of:

(1) $\left(\frac{1}{2}\right)^{-2}$ (2) $\left(\frac{2}{3}\right)^{-3}$ (3) $16^{-0.5}$
(4) $(36)^{-0.5}$ (5) $(4)^{1.5}$ (6) $\left(\frac{1}{4}\right)^{2.5}$

11   Find the value of $a^4 \times a^{-2} \times a^{\frac{1}{2}}$ when $a = 2$.

12   Write down the simplest forms of:

(1) $a^{\frac{1}{2}} \times a^{\frac{1}{3}}$ (2) $10^3 \times 10^{-\frac{1}{2}}$
(3) $a^3 \div (-a)^6$ (4) $-a^3 \div (-a)^5$
(5) $x^{2n} \div x^{-n}$ (6) $x^n \div x^{-n}$

13   Find the values of:

(1) $a^3 \times a^0$ (2) $a^0 \times 1$ (3) $a^0 \times 0$

14   Write down in 'standard form' the answers to the following:

(1) $(2.2 \times 10^5) \times (1.6 \times 10^4)$
(2) $(7.1 \times 10^3) \times (2.3 \times 10^3)$
(3) $(4.62 \times 10^5) \div (2.1 \times 10^3)$
(4) $(7.4 \times 10^6) \times (5 \times 10^{-4})$
(5) $(1.2 \times 10^{-3}) \times (2.1 \times 10^{-4})$

19

Logarithms

160 A SYSTEM OF INDICES

In the previous chapter you saw that by using the graph of 2^x you could, within the limits of the graph, express any number as a power of 2. This was confirmed algebraically.

For every number marked on the y-axis and indicated on the graph, there is a corresponding index which can be read on the x-axis. These constitute a system of indices by which numbers can be expressed as powers of a common basic number 2.

Similarly, by drawing graphs such as 3^x, 5^x, 10^x, numbers can be expressed as powers of 3, 5, or 10, or any other basic number.

Thus in all such cases it is possible to formulate systems of indices which, for any number A, would enable you to determine what power that number is of any other number B, which is called the base of the system.

This possibility of expressing any number as a power of any other number, and thus of the formation of a system of indices, as stated above, leads to practical results of great importance. Before calculators were developed around 1980, indices enabled people to carry out, easily and accurately, calculations which would otherwise have been laborious or even impossible. The fundamental ideas underlying this can be illustrated by means of a graph of powers similar to that drawn in Fig. 62. For this purpose 10 has been used as the base of the system and the graph of $y = 10^x$ has been drawn in Fig. 63.

As powers of 10 increase rapidly, it will be possible to employ only small values for x, if the curve is to be of any use. To obtain those powers we must use the rules for indices which were formulated in the previous chapter.

From arithmetic we know that $\sqrt{10} = 3.16$ app., i.e., $10^{\frac{1}{2}} = 3.16$. Then

$10^{0.25} = 10^{\frac{1}{4}} = (10^{\frac{1}{2}})^{\frac{1}{2}} = (3.16)^{\frac{1}{2}} = 1.78$ app. (by arithmetic).
$10^{0.75} = 10^{\frac{3}{4}} = 10^{\frac{1}{2}+\frac{1}{4}} = 10^{\frac{1}{2}} \times 10^{\frac{1}{4}} = 3.16 \times 1.78 = 5.62$ app.
$10^{0.125} = 10^{\frac{1}{8}} = (10^{\frac{1}{4}})^{\frac{1}{2}} = (1.78)^{\frac{1}{2}} = 1.33$ app

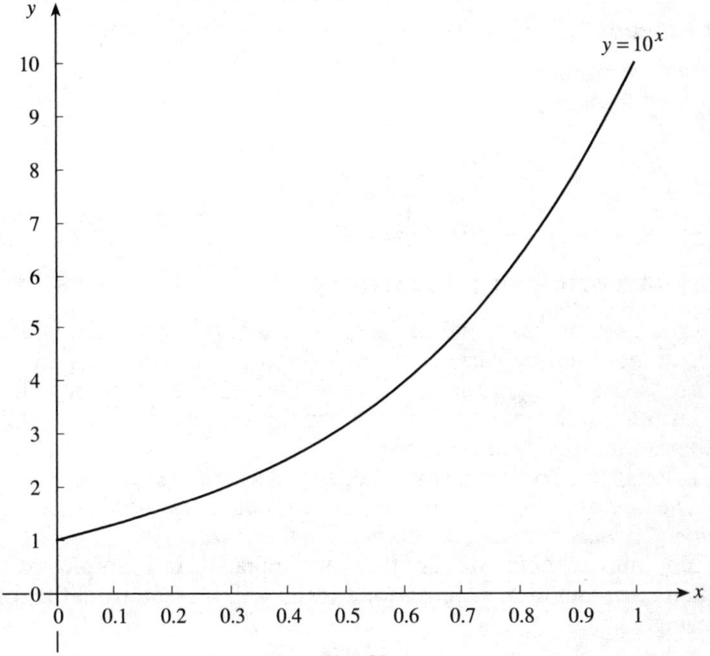

Fig. 63

In this way a table of values for the curve can be compiled as follows:

x	0	0.125	0.25	0.5	0.75	0.875	1
10^x	1	1.33	1.78	3.16	5.62	7.5	10

The resulting curve is shown in Fig. 63. The following examples illustrate the use that can be made of it in calculations.

■ *Example 1:* Find from the graph the value of 1.8×2.6.

From the graph

$$1.8 = 10^{0.26}$$
$$2.6 = 10^{0.42}$$
$$\therefore \quad 1.8 \times 2.6 = 10^{0.26} \times 10^{0.42}$$
$$= 10^{0.26+0.42} \text{ (first law of indices)}$$
$$= 10^{0.68}$$
$$= 4.6 \text{ from the graph.}$$

■ *Example 2:* Find $\sqrt[3]{9}$.

From the graph

$$9 = 10^{0.96}$$
$$\sqrt[3]{9} = 9^{\frac{1}{3}}$$
$$= (10^{0.96})^{\frac{1}{3}}$$
$$= 10^{0.32} \text{ (third law of indices)}$$
$$= 2.1 \text{ from the graph.}$$

161 A SYSTEM OF LOGARITHMS

Before 1980, tables of indices used to be constructed so that multiplications could be carried out by using addition, more or less in the way described in the previous section. The tables were called logarithm tables, or, more briefly, log tables, but they were really tables of indices by another name.

A logarithm to base 10 may be defined as follows:

The logarithm of a number to base 10 *is the index of the power to which* 10 *must be raised to produce the number.*

You may wonder why another, unfamiliar term is employed as a name for an index. One reason for this will be seen from the following:

Let n be any positive number.
Let x be its index to base 10.
Then $n = 10^x$.

This is in reality a formula. If it is required to 'change the subject of the formula' (see Section 57) and express x in terms of the other letters, there is a difficulty in doing this concisely. Using words we could write:

$$x = \text{index of power of } n \text{ to base 10.}$$

This is cumbersome, so the word 'logarithm'[1] is used, abbreviated to 'log' as follows:

$$x = \log_{10} n$$

the number indicating the base being inserted as shown.

If the base is e, you write $x = \log_e n$, or sometimes $x = \ln n$.

In this form x is expressed as a function of n, whereas in the form $n = 10^x$, n is expressed as a function of x.

You should be able to change readily from one form to another.

■ *Examples:*

(1) You saw in Section 160 that $5.62 = 10^{0.75}$.
     In log form this is $0.75 = \log_{10} 5.62$.

(2) $1024 = 2^{10}$.
     ∴ $\log_2 1024 = 10$.

(3) $1000 = 10^3$
     ∴ $\log_{10} 1000 = 3$.

(4) $81 = 3^4$.
     ∴ $\log_3 81 = 4$.

For many calculations 10 is the most suitable base for a system of logarithms, but in more advanced mathematics a different base is required (see Section 163).

162 RULES FOR THE USE OF LOGARITHMS

In using logarithms you must be guided by the laws which govern operations with them. Since logarithms are indices, these laws must be the same in principle as those of indices. These laws are given below; formal proofs are omitted. They follow directly from the corresponding index laws.

(1) **Logarithm of a product**
     The logarithm of the product of two or more numbers is equal to the sum of the logarithms of these numbers (see first law of indices).

     Thus if p and q are any numbers

     $$\log (p \times q) = \log p + \log q.$$

(2) **Logarithm of a quotient**
     The logarithm of p divided by q is equal to the logarithm of p diminished by the logarithm of q (see second law of indices).

     Thus $\log (p \div q) = \log p - \log q.$

[1] The choice of the word logarithm can be explained only by the history of the word. You could consult *A Short History of Mathematics*, by W.W.R. Ball.

(3) **Logarithm of a power**

The logarithm of a power of a number is equal to the logarithm of the number multiplied by the index of the power (see third law of indices).

Thus $$\log a^n = n \log a.$$

(4) **Logarithm of a root**

This is a special case of the above (3).

Thus $$\log \sqrt[n]{a} = \log a^{\frac{1}{n}}$$
$$= \frac{1}{n} \log a.$$

(5) **Logarithm of the base**

The logarithm to base b of the number b itself is 1. This is because any finite number b can be written as

$$b = b^1$$

so from section 161,

$$\log_b b = \text{the index of power } b \text{ to base } b$$
$$= 1$$

Hence from the example in section 161,

$$\log_{10} 10 = 1$$

(6) **Logarithm of a negative number**

The logarithm of a number less than zero does not, in reality, exist. To see this consider the following example in base 10. If n is any positive number so that $-n$ is negative, is it possible to find a number x, so that

$$-n = 10^x?$$

In other words, what is $x = \log_{10}(-n)$?

Clearly any power of 10 is always positive. It may be very large and positive, e.g., $10^{365.25}$. It may be very small, e.g., 10^{-747}, but even this number is still just greater than 0. This means that it is impossible to find a real number x such that. Hence it is not possible to find the logarithm of $x = \log_{10}(-n)$, i.e., the logarithm to base 10 of a negative number. The same is true for any other positive base.

(7) **The logarithm of 1 is always zero**

For any finite b, this can be seen to be true, since

$$1 = 2^0, \ 1 = 2.718^0, \ \dots \ 1 = 3^0, \ \dots \ 1 = 10^0, \ \dots \ 1 = b^0$$

hence by rule 3,

$$\log_2 1 = 0, \ \log_{2.718} 1 = 0, \ \dots \ \log_3 1 = 0, \ \dots \ \log_{10} 1 = 0, \ \dots \ \log_b 1 = 0$$

(8) The logarithm of a number is equal to the negative of the logarithm of its reciprocal

This follows from rule 2 with $p = 1$, and then from rule 7.

$$\log(1/q) = \log 1 - \log q \qquad \text{(rule 2)}$$
$$= \log 1 - \log q \qquad \text{(rule 7)}$$
$$\log(1/q) = -\log q$$

(9) The logarithm of a number between 0 and 1 is negative

This follows from rule 8 with $q > 1$. For example, if $q = 100$ and the base is 10, then

$$\log_{10}(100) = \log_{10}10^2 = 2 > 0$$

Hence we have from rule (8),

$$\log_{10}(1/100) = -\log_{10}100 = -2 < 0$$

163 CHANGE OF BASE OF A SYSTEM OF LOGARITHMS

Although logs calculated to base 10 were usually employed for calculations, in more advanced mathematics, as well as in engineering, the logs which naturally arise are calculated to a base which is given by the sequence

$$1 + \frac{1}{1} + \frac{1}{1.2} + \frac{1}{1.2.3} + \frac{1}{1.2.3.4} + \ldots \text{ to infinity.}$$

The sum of this sequence is denoted by e, and its value can be calculated to any required degree of accuracy by taking sufficient terms. To 5 places of decimals, $e = 2.718\,28$.

Logs calculated to this base are called Naperian logarithms, after Lord Napier, who discovered them in 1614, using this base. They are also called Natural logarithms or Hyperbolic logarithms.

The relations between the logs of numbers to different bases is found as follows:

Let n be any number.

Let a and b be two bases.

Suppose that logs to base b are known, and you require to find them to base a.

Let $\qquad \log_b n = y, \therefore n = b^y.$

Then $\qquad \log_a n = \log_a(b^y)$
$$= y \log_a b. \text{ (Section 162, rule 3.)}$$
$\therefore \qquad \log_a n = \log_b n \times \log_a b.$

Thus, knowing the log of a number to a base b, you can find its log to base a by multiplying, whatever the number, by $\log_a b$.

In the above result let $b = 10$ and $a = e$.

Then $\log_e n = \log_{10} n \times \log_e 10$.
In this result   let $n = e$
then $\log_e e = \log_{10} e \times \log_e 10$
                       but $\log_e e = 1$.
$\therefore$ $\log_{10} e \times \log_e 10 = 1$.

$\therefore$ $\log_e 10 = \dfrac{1}{\log_{10} e}$.

$\therefore$ in the rule $\log_e n = \log_{10} n \times \log_e 10$

you can write $\log_e n = \log_{10} n \times \dfrac{1}{\log_{10} e}$.

Thus both logs on the right-hand side are to base 10.

Now $\log_{10} e = 0.4343$

and $\log_{10} e = \dfrac{1}{0.4343} = 2.3026$.

Hence to change from base 10 to base e, you may use either of the following:

(1) $\log_e n = \log_{10} n \times 2.3026$
or   (2) $\log_e n = \log_{10} n \div 0.4343$.

■ *Example:* Find $\log_e 50$.

Using $\log_e 50 = \log_{10} 50 \times 2.3026$
you have $\log_e 50 = 1.6990 \times 2.3026 = 3.913$.

Here is a summary of the laws relating to logarithms.

EXERCISE 59

1 Without using a calculator, use the rules of logarithms to simplify:

    (a) $\log 27 \div \log 3$ (b) $(\log 16 - \log 2) \div \log 2$

2 Find the values of

    (1) $\log_e 4.6$ (2) $\log_e 0.062$

3 The insulating resistance, R, of a wire of length l is given by
$$R = \frac{0.42S}{l} \times \log_e \frac{d_2}{d_1} .$$
Find l when $S = 2000$,
$R = 0.44$, $d_2 = 0.3$, $d_1 = 0.16$.

4 In a calculation on the dryness of steam the following formula was used:
$$\frac{qL}{T} = \frac{q_1 L_1}{T_1} + \log_3 \frac{T_1}{T} .$$
Find q when $L_1 = 850$, $L = 1000$, $T_1 = 780$, $T = 650$, $q_1 = 1$.

20

Ratio and Proportion

164 MEANING OF A RATIO

There are two ways of comparing the magnitudes of two numbers:

(1) By subtraction. This operation states by how much one number is greater or less than the other. If the numbers are represented by a and b, the comparison is expressed by $a - b$.

(2) By division. By this means you find out what multiple or what part or parts one number is of the other.

The latter is called the **ratio** of the two numbers, and may be expressed by $a \div b$, or $\dfrac{a}{b}$, or in the special form $a : b$. Of these the fractional form is best suited for manipulation.

165 RATIO OF TWO QUANTITIES

The magnitude of two quantities of the same kind, such as two lengths, weights, sums of money, etc., may be compared by means of a ratio. To do this the measures of the two quantities are expressed in terms of the same unit by numbers. The ratio of these two numbers expresses the ratio of the quantities.

Thus, the ratio of two distances which are respectively a metres and b metres would be $\dfrac{a}{b}$ or $a : b$.

The ratio would be unaltered in value if the quantities were expressed in other units provided both quantities were expressed in the same units.

Thus the ratio of 3 hours to 2 hours is the same as the ratio of 9 metres to 6 metres or 108° to 72°.

This is obvious from the consideration that, as a ratio can be expressed by a fraction, it can be manipulated in the same ways as fractions.

Thus

$$\frac{a}{b} = \frac{a \times m}{b \times m}$$

and

$$\frac{a}{b} = \frac{a \div m}{b \div m} \,.$$

A ratio is always a number, either an integer or a fraction (vulgar or decimal), and is not expressed in terms of any particular unit provided both quantities are in the same units.

166 PROPORTION

If four numbers a, b, c, d are related so that the ratios $\dfrac{a}{b}$ and $\dfrac{c}{d}$ are equal, the numbers are said to be in proportion.

It follows from the definition of a ratio that a divided by b must represent the ratio of two quantities of the same kind, while c divided by d must also represent the ratio of two quantities of the same kind, though not necessarily of the same kind as a and b.

Thus a and b might represent the measures of two weights, while c and d might represent the measures of two costs.

Continued proportion

If a sequence of numbers a, b, c, d... is such that

$$\frac{a}{b} = \frac{b}{c} = \frac{c}{d} = \cdots$$

then these numbers are said to be in continued proportion.

Thus in the sequence of numbers 2, 6, 18, 54 ... the ratios

$$\frac{2}{6} = \frac{6}{18} = \frac{18}{54} \cdots$$

Mean proportional   If a, b, c are numbers such that

$$\frac{a}{b} = \frac{b}{c} \,,$$

b is called a mean proportional between a and c.

Then $b^2 = ac$

and $b = \sqrt{ac}.$

In this way you can find a mean proportional between any two numbers.

167 Theorems on ratio and proportion

You should note the following results:

(1) Let $\dfrac{a}{b} = \dfrac{c}{d}.$

Then $\dfrac{a}{b} \times bd = \dfrac{c}{d} \times bd.$

$\therefore$ $ad = bc.$

(2) Let $\dfrac{a}{b} = \dfrac{c}{d}.$

Then from the meaning of a ratio

$$\frac{b}{a} = \frac{d}{c}.$$

(3) Let $\dfrac{a}{b} = \dfrac{c}{d}.$

Then $ad = bc$

$\therefore$ $\dfrac{ad}{cd} = \dfrac{bc}{cd}.$

$\therefore$ $\dfrac{a}{c} = \dfrac{b}{d}.$

168 An illustration from geometry

The following illustration from geometry is given as being of great importance. For a proof of it and for a geometrical treatment of ratio and proportion you are referred to any textbook on geometry.

Similar triangles. Triangles which have their corresponding angles equal are called **similar**. In such triangles,

the ratios of corresponding sides are equal.

In Fig. 64 the triangles ABC, A'B'C' are similar – i.e.,

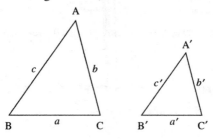

Fig. 64

If ∠A denotes the angle at vertex A etc, then ∠A = ∠A′, ∠B = ∠B′, ∠C = ∠C′. Corresponding sides are those which are opposite to equal angles.

Denoting the lengths of the sides opposite to these by a, b, c, $a′$, $b′$, $c′$ as shown, then by the theorem stated above

$$\frac{a}{a′} = \frac{b}{b′} = \frac{c}{c′} .$$

It follows from theorem (3) of Section 167

$$\frac{a}{b} = \frac{a′}{b′} , \frac{b}{c} = \frac{b′}{c′} \text{ and } \frac{a}{c} = \frac{a′}{c′} ,$$

i.e., the ratios of the pairs of sides containing the equal angles are equal. Each pair of equal ratios gives a set of numbers in proportion.

169 CONSTANT RATIOS

Let AOB be any angle (Fig. 65).

On the arm OA take points P, Q, R ... and draw PK, QL, RM perpendicular to the other arm.

Then the △s OPK, OQL, ORM are equiangular and similar.

∴ by the theorem of Section 168

$$\frac{PK}{OK} = \frac{QL}{OL} = \frac{RM}{OM}$$

Clearly any number of points, such as P, Q, R, can be taken and the ratio of all pairs of sides such as the above are equal. You can therefore say that for the angle AOB all such ratios are constant in value.

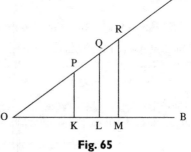

Fig. 65

This constant ratio is called in trigonometry the tangent of the angle AOB. It is abbreviated to tan AOB – i.e.,

$$\frac{PK}{OK} = \frac{QL}{OL} = \frac{RM}{OM} = \tan AOB.$$

A similar result is true for any other angle.

∴ every angle has its own tangent or constant ratio, by which it can be identified.

Referring to Sections 82 and 83 it will be seen that the gradient of a straight line, represented by m, is the tangent of the angle made by the line with the x-axis.

In Fig. 35, for example, the ratio $\dfrac{PQ}{OQ}$ is constant for every point on the line, and is the tangent of the angle POQ.

In the general equation $y = mx + b$, m represents the tangent of the angle made with the x-axis.

170 EXAMPLES OF EQUAL RATIOS

Examples of equal ratios frequently occur in mathematics, and the following theorem, in different forms, is sometimes useful.

Let $\dfrac{a}{b} = \dfrac{c}{d} = \dfrac{e}{f}$ be equal ratios (there may be any number of them).

Let k represent their common value.

Then
$$\frac{a}{b} = \frac{c}{d} = \frac{e}{f} = k.$$

∴
$$a = bk$$
$$c = dk$$
$$e = fk.$$

These results make possible various manipulations.

For example, by addition

$$a + c + e = bk + dk + fk$$
$$= k(b + d + f).$$

∴
$$\frac{a + c + e}{b + d + f} = k,$$

and is therefore equal to each of the original ratios. This can be varied in many ways.

For example, by multiplication of the three equations above:

$$2a = 2bk$$
$$3c = 3dk$$
$$7e = 7fk.$$

∴ by addition

$$2a + 3c - 7e = 2bk + 3dk - 7fk$$
$$= k(2b + 3d - 7f).$$

∴ $$\frac{2a + 3c - 7e}{2b + 3d - 7f} = k = \text{each of the original ratios.}$$

EXERCISE 60

1 Write down the following ratios:

   (a) £a to b pence.
   (b) p hours to q minutes r seconds.

2 Write down the ratio of 'a km an hour' to 'a m per second'.

3 If the ratio $a : b$ is equal to the ratio $5 : 8$, find the numerical values of the following ratios:

   (1) $\dfrac{1}{a} : \dfrac{1}{b}$ (2) $a^2 : b^2$ (3) $2a : 3b$

4 (a) Two numbers are in the ratio of $4 : 5$. If the first is 28, what is the second?

   (b) Two numbers are in the ratio of $a : b$. If the first is x, what is the second?

   (c) Two numbers are in the ratio of $1 : x$. If the second is a, what is the first?

5 If the ratio $\dfrac{a}{b + c}$ is equal to the ratio $\dfrac{c}{x}$, find x.

6 A rectangle of area A mm^2 is divided into two parts in the ratio $p : q$. Find expressions for their areas.

7 A piece of metal of mass a kg is divided into two parts in the ratio $x : y$. What are the masses of the parts?

8 If $\dfrac{1}{a}$, $\dfrac{1}{b}$, $\dfrac{1}{c}$, $\dfrac{1}{x}$ are four numbers in proportion, find x.

9 What number must be added to each term of the ratio $11 : 15$, so that it becomes the ratio $7 : 8$?

10 Find the mean proportional between:

(1) ab and bc (2) $8a^2$ and $2b^2$

(3) $a(a + b)$ and $b(a + b)$

11 If $\dfrac{a}{b} = \dfrac{5}{2}$, find the value of $\dfrac{a+b}{a-b}$.

12 Find the ratio of $\dfrac{a}{b}$ when

(1) $3a = 7b$ (2) $\dfrac{3}{a} = \dfrac{7}{b}$

(3) $16a^2 = 25b^2$

13 (1) What is the result when 270 is increased in the ratio $5 : 3$?

(2) What is the result of increasing $\dfrac{a}{b}$ in the ratio $6 : 5$?

(3) What is the result of increasing $\frac{4}{3}$ in the ratio $a : b$?

14 If a, x, y, b are in continued proportion, find x and y in terms of a and b.

21

Variation

171 DIRECT VARIATION

In Chapter 16 examples are given of a variable quantity the value of which depended on the value of another variable, and is called a function of it. There are very many different forms which this variation may take, and in this chapter one of the most important of them is examined. Here is an example.

If a man is paid at a certain rate for every hour that he works, the total amount he earns over any period depends on the number of hours he works. If he doubles the number of hours he works, his earnings will be doubled. Generally the ratio of the amounts he earns in any two periods is equal to the ratio of the number of hours worked in the periods.

If T_1 and T_2 represent the number of hours worked in two periods and if W_1 and W_2 represent the amounts of wages earned in them,

then $$\frac{W_1}{W_2} = \frac{T_1}{T_2}.$$

These four quantities are in proportion (see Section 166).

Hence, when the relations between the two quantities can be expressed in this way you say that

The wages are proportional to the time worked or the wages vary directly as the time.

The wages are a function of the time, and the words 'proportional to' or 'vary directly as' are used to define the exact functional relations which exist between the two quantities.

Direct variation may be defined as follows:

If two quantities Y and X are so related that $\dfrac{Y_1}{X_1} = \dfrac{Y_2}{X_2}$ *, where X_1*

and X_2 are any two values of X and Y_1 and Y_2 are corresponding values of Y, then Y is said to be proportional to X, or Y varies directly as X.

In order to discover whether or not one quantity varies directly as another, this simple test can be applied.

If one quantity is doubled, is the other doubled in consequence? Or, more precisely, if one quantity is altered in a certain ratio, is the other altered in the same ratio?

The sign $\propto$ is used to denote direct variation. Thus in the case above you would write $Y \propto X$.

172 EXAMPLES OF DIRECT VARIATION

(1) The distance travelled by a car moving with uniform speed varies directly as the time.
(2) The mass of an amount of water is proportional to the volume.
(3) The circumference of a circle varies directly as the diameter.
(4) The electrical resistance of a wire varies directly as the length.

173 THE CONSTANT OF VARIATION

If $y \propto x$ then $y = kx$, where k is a constant.

In Section 170 the common value of a number of equal ratios was represented by k, a constant number for these ratios.

When one quantity varies directly as another, then you have seen that the ratio of corresponding pairs of values of the variables is constant. Consequently, as in Section 170, this constant is usually represented by k, and, using it, the relation between the quantities can be expressed as a formula.

For instance, in Example (1) of Section 172 it was stated that for a body moving with uniform speed, distance moved varies directly as time.

Let s represent any distance travelled.

Let t represent the corresponding time.

Then, since s varies directly as t, any ratio of corresponding values of these is constant.

Let k represent this constant.

Then by definition $\dfrac{s}{t} = k.$

$\therefore$ $s = kt.$

This is the form of the law connecting s and t, but it cannot be of much use until the value of k is known for this particular case.

To find this, you must know a pair of corresponding values of s and t. Thus if you are told that the car moves 40 m in 2.5 s, then, substituting the values of s and t, you get

$$40 = k \times 2.5.$$
$\therefore$ $k = 40 \div 2.5$
$$= 16.$$

$\therefore$ the law connecting s and t for this particular velocity is

$$s = 16t.$$

174 GRAPHICAL REPRESENTATION

If x and y represent two variables such that $y \propto x$, then, as shown in Section 173, $y = kx$.

The form of this equation is the same as $y = mx$, the graph of which was shown in Section 82 to be a straight line passing through the origin, m being the gradient of the line.

$\therefore y = kx$ represents a straight line of gradient equal to k.

Consequently the graphical representation of the variation of two quantities where one varies directly as the other is a straight line passing through the origin.

175 TO FIND THE LAW CONNECTING TWO VARIABLES

The engineer and the scientist frequently require to know the law connecting two variables, corresponding values of which have been found by experiments. The law may assume many forms, but in some cases there may be reason to suppose that one of the quantities varies directly as the other – i.e., the law may be of the form $y = kx$.

From the results of the experiment we can proceed to determine whether:

(1) the law is one of direct variation;
(2) if it is, and the law is of the form $y = kx$, the value of k must be found.

To determine (1) the results are plotted. Then:

(*a*) if the graph is a straight line, the law is one of direct variation;

(*b*) if the straight line passes through the origin, the equation is of the form $y = kx$.

Then k has to be found.

Graphically, k can be found as in Section 169 by finding the tangent of the angle made with the x-axis.

Algebraically, as shown in Section 173, a pair of corresponding values of x and y is chosen from the graph. These are substituted in $y = kx$, and so k is determined.

176 WORKED EXAMPLE

A spiral spring was extended by hanging various weights from it.

The amounts of extension of the spring for different weights were observed and tabulated as follows:

Weight in newtons	0	0.1	0.2	0.3	0.4	0.5
Extension in cm	0	0.15	0.3	0.44	0.6	0.75

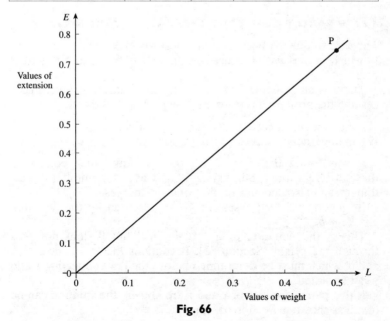

Fig. 66

From these results discover the law connecting the attached weight and the extension of the spring.

The graph resulting from plotting these results is shown in Fig. 66. It is a straight line passing through the origin though one of the points, corresponding to a weight of 0.3 N, lies slightly off the line. This is to be expected in experimental results. Also the line passes through the origin.

∴ as shown in Section 175, the law connecting the weight and extension is one of direct variation – i.e.,

The extension varies directly as the attached load.

Let E = extension in centimetres.
Let L = load in newtons.

Then $E \propto L$ and $E = kL$.
To find k a pair of values is taken, at the point P where $E \propto 0.75$ and $L = 0.5$.

Substituting these $0.75 = k \times 0.5$
whence $k = 1.5$.
∴ the law is $E = 1.5L$.

177 y PARTLY CONSTANT AND PARTLY VARYING AS x

The case frequently occurs, in practical work, of a variable quantity which in part varies directly as another quantity and in part is constant.

There is an example of this in Section 77 in the problem concerning the profit of a restaurant. The profit depends on

(1) the number of customers which is variable;
(2) the overhead charges which are constant.

It was found that $y = ax - b$ was the law which connected these, where y, the profit, varies directly as x, the number of customers, and also depends on the constant charges b.

In general all such cases can be represented by the equation $y = kx + b$.

This is the equation of a straight line, but it does not pass through the origin (Section 83). It contains the two constants k and b, which must be determined before the law connecting x and y can be stated.

If two pairs of values of x and y are known, the solution can be found as shown in Section 65, example 2.

In practical work pairs of values are found by experiment. The worked example which follows shows the method of procedure in such cases.

178 Worked example

When two voltmeters are compared they have readings corresponding to C and K below.

C	1.9	2.75	3.8	4.8	5.8
K	5.75	8.3	11.2	14	16.8

It is thought that C and K are connected by a law of the form

$$K = mC + b.$$

Test this by plotting the points and find the values of m and b.

The law $K = mC + b$ is linear – i.e., it is in the general form of the equation of a straight line. To test them you must find if the plotted points lie on a straight line.

Comparison of the two sets of values suggests that the scale units for C on the x-axis should be larger than those for K on the y-axis.

When the points are plotted you can see that they lie approximately on a straight line, slight deviations being due to experimental errors. When this line is drawn as evenly as possible, it appears as in Fig. 67.

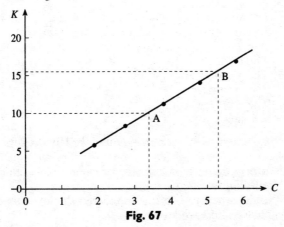

Fig. 67

Two suitable points, A and B, are selected on the line, and their coordinates are as follows:

For A, $C = 3.4$, $K = 10$.

For B, $C = 5.3$, $K = 15.5$.

These are to satisfy the equation

$$K = mC + b$$

Substituting you get:

$$10 = 3.4m + b \qquad (1)$$
$$15.5 = 5.3m + b \qquad (2)$$

Subtracting, $\quad 1.9m = 5.5$

∴ $$m = \frac{5.5}{1.9} = 2.9 \text{ approx.}$$

Substituting for m in (1),

$$10 = (3.4 \times 2.9) + b.$$
∴ $$b = 10 - 9.86$$
$$= 0.14.$$

∴ the law is $\quad K = 2.9C + 0.14.$

EXERCISE 61

1 The following are examples in which the value of one quantity depends on another. State in each case whether or not it is a case of direct variation.

   (*a*) Distance and time when a man runs the 1000 m race.
   (*b*) Interest and time when money bears interest at a fixed rate.
   (*c*) The logarithm of a number and the number itself.
   (*d*) The *y*- coordinate of a point on a straight line and the *x*- coordinate.
   (*e*) The cost of running a school and the number of scholars.

2 If $y = kx$ and $y = 8$ when $x = 7$, find k. Hence find y when $x = 40$.

3 If y is proportional to x and $y = 10$ when $x = 4$, find y when $x = 15$; also find x when $y = 8.4$.

4 If $y \propto x$, and when $y = 16.5$, $x = 3.5$, find the law connecting x and y. Hence find x when $y = 21$.

5 The distances travelled by a body from rest were as follows:

Time (s)	1	2	3	4
Distance (m)	3.2	6.4	9.6	12.8

Plot these and find if distance varies directly as time. If it does, find:

(1) the law connecting time (t) and distance (s);
(2) the distance passed over in 2.8 s.

6 The extension (E) of a spring varies directly as the force (W) by which it is stretched. A certain spring extended 2.4 when stretched by a weight of 4.5. Find:

(1) the law which connects them;
(2) the extension due to a weight of 7.

7 The law connecting two variables x and y is of the form $y = kx + b$, where k and b are constants. When $x = 10$, $y = 11$, and when $x = 18$, $y = 15$. Find k and b and state the law.

8 In a certain machine the law connecting the applied force (E) and the load (W) was of the form $E = aW + b$, where a and b are constants. When $W = 20$, $E = 1.4$, and when $W = 30$, $E = 2$. Find the law.

9 In experiments to determine the friction (F N) between two metallic surfaces when the load is W N, the following results were found:

W	3	5	7	10	12
F	0.62	1.5	2.4	3.6	4.4

Assuming W and F to be connected by a law of the form $F = aW + b$, find the law by drawing the average straight line between the points.

179 y VARIES AS THE SQUARE OF x — I.E., $y \propto x^2$

If the sides of a square are doubled, the area is not doubled, but multiplied by four; if the sides are trebled the area is multiplied by nine. The area of a square varies directly as the square of the length of its sides.

The ratio of the area of a circle to the square of the length of its radius can be shown experimentally to be constant. If A = the area and r = radius, then $\dfrac{A}{r^2}$ is constant for all circles. This constant is represented by the special symbol π, and its value is approximately 3.1416.

∴ the area of a circle varies directly as the square of the length of its radius.

Students of mechanics will know that the distance passed over by a body moving with uniformly increasing velocity is proportional to the square of the time. A special case is a falling body.

If s = the distance fallen in metres and t = time taken in seconds, then $s \propto t^2$ and $s = kt^2$.

Experiments show that $k = \frac{1}{2}g$, where g is an absolute constant whose value is approximately 9.81 m/s^2.

The graphical representation of $y = kx^2$ is that of a quadratic function (see Chapter 16). For different positive values of k, the graph is a parabola, symmetrical about the y-axis and having its lowest point at the origin (Section 125).

180 y VARIES AS THE CUBE OF x – I.E., $y \propto x^3$

If the edge of a cube is doubled, the volume is multiplied by eight. The volume varies directly as the cube of the length of an edge.

The volume of a sphere also varies as the cube of the radius.

If V = the volume and r = the length of the radius, then $V \propto r^3$ and $V = kr^3$.

It can be shown that $k = \frac{4}{3}\pi$.

∴ $$V = \tfrac{4}{3}\pi r^3.$$

The graph of $y = x^3$ is a curve, as shown in Fig. 68. It is called a cubic curve.

You should make a table of values and draw the curve, or draw it on a graphics calculator.

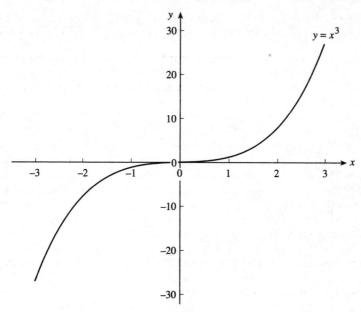

Fig. 68

181 y VARIES AS $\sqrt{x}$ OR $x^{\frac{1}{2}}$, I.E., $y \propto \sqrt{x}$

This form of variation, besides arising in various physical examples, may also be regarded as the inverse of $y = x^2$.

Since
$$y = kx^2$$
$$x^2 = \frac{y}{k}.$$

$\therefore$
$$x = \sqrt{\frac{1}{k}} \times \sqrt{y}.$$

Since $\sqrt{\dfrac{1}{k}}$ is a constant, then $x \propto \sqrt{y}$.

The graph of $y = \sqrt{x}$ is part of a parabola, as in Fig. 69. If the graph of $y = -\sqrt{x}$ were also drawn, it would be the same shape as Fig. 50, but it would be symmetrical about the x-axis instead of the y-axis.

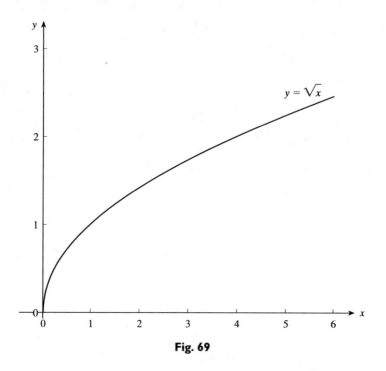

Fig. 69

182 INVERSE VARIATION: $y \propto \dfrac{1}{x}$

Let x and y be two numbers such that their product is constant – i.e., $xy = k$,

then $$y = \frac{k}{x} \text{ or } x = \frac{k}{y}.$$

Each quantity can be expressed in terms of the reciprocal or inverse of the other.

Many examples of this occur in mathematics. Here is a simple case.

Let x and y be the sides of a rectangle of area 60 m^2, then $xy = 60$.

The lengths of the sides may be varied in very many ways, their product always being equal to 60. If x is increased, y will be decreased, and vice versa. If x is doubled, y will be halved. In general, if x is changed in a given ratio, y will be changed in the inverse ratio.

Hence y is said to vary inversely as x – i.e., $y \propto \dfrac{1}{x}$.

Hence $$y = \frac{k}{x} .$$

Here are some examples of inverse variation.

(1) Time to travel a given distance varies inversely as the speed. If the speed is doubled, the time is halved.

(2) The volume of a fixed mass of gas varies inversely as the pressure on it, the temperature remaining constant.
If p = the pressure and v = the volume.

$$p \propto \frac{1}{v} \text{ or } p = \frac{k}{v} .$$

(3) The electrical resistance of a wire of given length and material to the passage of a current through it varies inversely as the area of the cross-section of the wire.
If R = the resistance,
and A = the area of the cross section,

then $$R \propto \frac{1}{A} \text{ or } R = \frac{k}{A} .$$

183 GRAPH OF $y = \dfrac{k}{x}$

In its simplest form when $k = 1$, the equation becomes $y = \dfrac{1}{x}$.

The graph of this function presents some new difficulties which will be apparent on drawing the curve. The following is a table of values formed in the usual way:

x	$\frac{1}{4}$	$\frac{1}{2}$	1	2	3	4
y	4	2	1	$\frac{1}{2}$	$\frac{1}{3}$	$\frac{1}{4}$

A similar set of values can be tabulated for negative values of x, the corresponding values of y being negative.

The curve, which is called a hyperbola, is shown in Fig. 70. It consists of two branches, alike in shape, one for +ve values and

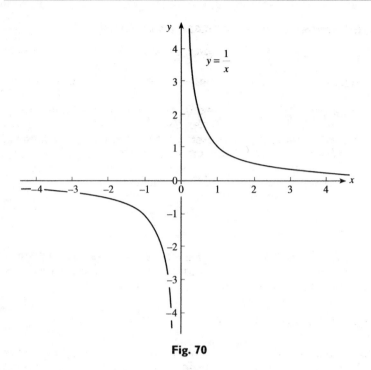

Fig. 70

the other for −ve values of x. Note the following important features of this curve.

(1) As x increases, y or $\dfrac{1}{x}$ decreases. When x becomes very great,

y becomes very small in magnitude, and the curve approaches very close to the x-axis.

(2) As x decreases, y or $\dfrac{1}{x}$ increases. When x is very small, y is

very large in magnitude, and the curve approaches very close to the y-axis (for $x > 0$, $y > 0$, for $x < 0$, $y < 0$). Both these features of the curve are repeated for −ve values of x. It may be noted that the curve is symmetrical (i) about the line through the origin making 45° with the x-axis – i.e., the line $y = x$; (ii) about a line through the origin at right angles to this – i.e., the line $y = -x$ (see Sections 82 and 83).

For other values of k the curve of $y = \dfrac{k}{x}$ is always a hyperbola.

184 OTHER FORMS OF INVERSE VARIATION

(1) One quantity may vary inversely as the square of another quantity – i.e., $y \propto \dfrac{1}{x^2}$, whence $y = \dfrac{k}{x^2}$. In electricity, for example, the force between two magnetic poles varies inversely as the square of the distance between them. Many other physical laws involve this form of variation.

(2) Another form of variation is that in which one quantity varies inversely as the cube of another – i.e., $y \propto \dfrac{1}{x^3}$ and $y = \dfrac{k}{x^3}$.

Generalising, y may vary directly or inversely as any power of x – i.e., $y \propto x^n$ or $y \propto \dfrac{1}{x^n}$.

In all cases of direct variation the same method, the introduction of the constant k, is followed and the evaluation of k proceeds on the same lines.

185 WORKED EXAMPLES

■ *Example 1:* If y varies as the cube root of x, and if $y = 3$ when $x = 64$, find the formula connecting the variables. Hence find x when $y = \frac{15}{4}$.

(1) Since

$$y \propto x^{\frac{1}{3}}$$
$$y = kx^{\frac{1}{3}}.$$
$$\therefore \qquad k = y \div x^{\frac{1}{3}}.$$

Substituting the given values

$$y = 3 \div 64^{\frac{1}{3}}$$
$$= 3 \div 4 \text{ or } \tfrac{3}{4}.$$

∴ the law is
$$y = \tfrac{3}{4} \sqrt[3]{x}.$$

(2) When $y = \frac{15}{4}$, you have

$$\tfrac{15}{4} = \tfrac{3}{4} \sqrt[3]{x}.$$
$$\therefore \qquad \sqrt[3]{x} = \tfrac{15}{4} \div \tfrac{3}{4}.$$
$$= 5.$$
$$\therefore \qquad x = 5^3 = 125.$$

■ *Example 2:* The time of vibration of a simple pendulum varies as the square root of its length. If the length of a pendulum which beats every 1 second is 1 m, what will be the time of vibration if its length is increased by 6 cm?

Let l = length of the pendulum in metres and
let t = the time of variation.

Then $\qquad\qquad t \propto \sqrt{l}$ and $t = k\sqrt{l}$.

When $l = 1.00$, $t = 1.00$ and $1 = k\sqrt{1}$, $\therefore k = 1$.

When $l = 1.06$, $t = k\sqrt{l} = 1\sqrt{1.06} = 1.03$.

$\therefore \qquad\qquad\qquad t = 1.03$ s.

EXERCISE 62

1 If y is proportional to x^2 and when $x = 15$, $y = 200$, find the equation connecting x and y. Find y when $x = 8.5$.

2 If $y \propto x^3$ and $x = 2$ when $y = 2$, find the law connecting x and y. Then find y when $x = 3$.

3 If $y \propto \dfrac{1}{x^2}$, fill up the blanks in the following table:

x	1	2	3	
y			$\frac{4}{3}$	$\frac{1}{3}$

4 If $y \propto \sqrt{x}$ and if $y = 3.5$ when $x = 4$, express y in terms of x. What is y when $x = 25$?

5 If $y \propto x^3$ and if $y = 6$ when $x = 4$, find the value of y when $x = 16$. Also find x when $y = 3$.

6 If y is inversely proportional to x, and $x = 5$ when $y = 6$, find the law connecting x and y and find x when $y = 20$.

7 If $y \propto \dfrac{1}{x}$, fill up the gaps in the following table:

x		1.2	8	
y	6		1.5	0.8

8 The force F which acts between two magnetic poles is inversely proportional to the square of the distance (d) between them. Express this as a formula if $F = 120$ when $d = 4$.

9 The distance through which a heavy body falls from rest varies as the square of the time taken. A lead ball falls through 490 m in 10 s. How long will it take to fall through 78.4 m?

10 The intensity of illumination given by a projector varies inversely as the square of the distance of its lamp from the screen. If the projector is 20 m from the screen, where must it be placed so that the illumination is 4 times as great?

11 If the distances of an object and of its image, formed by a mirror, are measured from a certain point, it is found that the sum of the distance varies as their product. If the image distance is 120 mm when the object distance is 300 mm, calculate the image distance when the object distance is 540 mm.

12 The square of the speed of a particle varies as the cube of its distance from a fixed point. If this distance is increased by 1.2 per cent, what is the approximate percentage increase in the speed?

13 A clock keeps accurate time at 10°C, but gains as the temperature falls and vice-versa, the rate of gain or loss varying as the square of the number of °C between the actual temperature and 10°C. If it gains 2 seconds per day when the temperature is 5°C, how much does it lose (to the nearest second) in 4 days when the temperature is 42°C?

14 If $y \propto x^{1.4}$ and if $y = 354.5$ when $x = 15$, find the law connecting y and x.

186 FUNCTIONS OF MORE THAN ONE VARIABLE

■ *Example 1:* It is proved in elementary geometry that the area of a triangle is given by the formula

$$A = \tfrac{1}{2}bh$$

where A = the area of the triangle
b = the length of a base
h = the corresponding height.

Both b and h are variables, and the value of A depends on them both.

∴ A is a function of the two variables b and h.

In any triangle let the height remain constant, but the base variable, then if the base is doubled, the area will be doubled. If the base remain constant and the height is trebled, the area will be trebled.

Now suppose both base and height to vary; let the base be doubled and the height trebled, then the area will be 2×3 – i.e., 6 times greater.

You can infer, then, that if both base and height vary, the area varies as the product of base and height –i.e.,

$$A \propto b \times h.$$
$$\therefore \qquad A = kbh.$$

From geometrical considerations you know that in this case $k = \frac{1}{2}$.

■ *Example 2:* In Section 182 it was stated that the volume of a given mass of gas, at a constant temperature, varies inversely as the pressure on it. It can also be shown by experiment that if the pressure is kept constant and the temperature varied, then the volume varies directly as the absolute temperature. If both temperature and pressure vary, then the volume varies directly as the absolute temperature and inversely as the pressure.

Let v = volume
T = absolute temperature
p = pressure

then
$$v \propto \frac{T}{p}.$$

$$\therefore \qquad v = k \times \frac{T}{p}.$$

■ *Example 3:* If an electric current passes through a wire, it encounters resistance. This resistance varies as the length of the wire (Section 172), and in a wire of given length – i.e., constant – it varies inversely as the cross section of the wire (Section 182).

$\therefore$ if R = the resistance
l = the length
A = the cross section

then R varies directly as l and inversely as A.

Thus
$$R \propto \frac{l}{A}$$

and
$$R = k \times \frac{l}{A}.$$

187 JOINT VARIATION

The variation of a quantity due to two or more variables is sometimes called joint variation, and the quantity is said to vary jointly
as their product.

In dealing with problems involving joint variation, the same
procedure with regard to the constant of variation and its determination is followed as in the previous case. The following examples
will serve to illustrate it.

188 WORKED EXAMPLES

■ *Example 1:* A quantity represented by y varies directly as x and
inversely as z^3. It is known that when $x = 15$, $z = 12$ and $y = \frac{1}{36}$.
Find the law connecting the quantities.

You are given that $\quad y \propto \dfrac{x}{z^3}$.

$\therefore \qquad\qquad\qquad y = k \times \dfrac{x}{z^3}$.

Substituting the given values

$$\frac{1}{36} = k \times \frac{15}{12^3}$$.

$\therefore \qquad\qquad k = \dfrac{1}{36} \div \dfrac{15}{12^3} = \dfrac{16}{5}$.

$\therefore$ the law is $\qquad y = \dfrac{16}{5} \dfrac{x}{z^3}$.

■ *Example 2:* The force between two magnetic poles varies jointly as the product of their strength and inversely as the square of
the distance between them. If two poles of strength 8 and 6 units
repel one another with a force of 3 N when placed 4 m apart, with
what force will two poles whose strengths are 5 and 9 units repel
one another when 2 m apart?

Let $\qquad F =$ the force
$m_1, m_2 =$ the pole strengths
$d =$ the distance apart

then $\qquad\qquad F \propto \dfrac{m_1 m_2}{d^2}$.

$\therefore \qquad\qquad F = k \times \dfrac{m_1 m_2}{d^2}$.

Substituting the given values

$$3 = k \times \frac{8 \times 6}{4^2}.$$

$$\therefore \qquad k = \frac{3 \times 4^2}{8 \times 6} = 1.$$

$$\therefore \qquad F = \frac{m_1 m_2}{d^2}.$$

In the second case $F = \dfrac{5 \times 9}{4} = 11.25$ N.

EXERCISE 63

1   Express the following statements in the form of equations:

(a) y varies jointly as x and z.

(b) y varies directly as x and inversely as the square of z.

(c) y varies directly as the square root of x and inversely as z.

(d) The volume (V) of a cylinder varies jointly as the height (h) and the square of the radius of the base (r).

(e) The weight (W) which can be carried safely by a beam varies inversely as the length (l), directly as the breadth (b) and directly as the square of the depth (d).

(f) y varies directly as the square of x and inversely as the cube root of z.

2   If y varies directly as x and inversely as z, and if $y = 10$ when $x = 8$ and $z = 5$, find the law connecting x, y and z. Also find y when $x = 6$ and $z = 2.5$.

3   If y varies jointly as x and z^2, and if $y = 13\frac{1}{3}$ when $x = 2.5$ and $z = \frac{4}{3}$, find the law connecting the variables. Also find x when $z = \frac{3}{2}$ and $y = 54$.

4   y varies directly as x^2 and inversely as $\sqrt{z}$. When $x = 8$ and $z = 25$, $y = 16$. Find y when $x = 5$ and $z = 9$.

5   The load that a beam of given depth can carry is directly proportional to the breadth and inversely proportional to the length. If a beam of length 7 m and width 175 mm can support a load of 4 tonnes, what load can be supported by a beam of the same material 5 m long and 250 mm wide?

6 If z varies as x^2 and inversely as y^2, and if $z = 4$ when $x = 8$ and $y = -0.5$, find z when $x = -2$ and $y = 0.25$.

7 The number of heat units (H) generated by an electric current varies directly as the time (t) and the square of the voltage (E) and inversely as the resistance (R). If $H = 60$, when $t = 1$, $E = 100$ and $R = 40$ find the law connecting them.

Also find:

   (1) the value of H when $E = 200$, $R = 120$ and $t = 300$;
   (2) the value of t when $E = 120$, $R = 90$ and $H = 5760$.

8 The pressure in a test apparatus varies directly as the absolute temperature and inversely as the volume of gas. If the temperature increases by 2 per cent and the volume decreases by 2 per cent, what is the change in the pressure?

22

The Determination of Laws

189 LAWS WHICH ARE NOT LINEAR

In the preceding chapter you considered the determinations of laws which were linear, and which were arrived at by using experimental data.

But such laws are not always represented by straight lines. They may involve powers of a variable such as were considered in Section 179 and onwards. In these cases, when the results of the experiments are plotted, they will lie on a portion of a curve which might be one of those illustrated in the preceding chapter or of many others. In practice, when only a small portion of the curve can be drawn, it is impossible to identify what curve it is.

There are two devices, however, by means of which a straight line can be obtained instead of a curve. The identification can then be made by the methods previously considered.

190 $y = ax^n + b$. PLOTTING AGAINST A POWER OF A NUMBER

If the law which you require to find is of the form $y = ax^n + b$ or $y = ax^n$, where n is known, then plot the corresponding values of y and x^n instead of y and x. The resulting graph will be a straight line.

Consider as an example $y = 2x^2$.

The graph of this function was found in Section 125 to be a parabola. The values given to x were $0, 1, 2, 3 \ldots$

The corresponding tabulated values of y and x^2 will be

x^2	0	1	2	3	4
y	0	2	4	6	8

and so on. The values of x are not shown in the table.

The resulting graph is a straight line passing through the origin as shown in Fig. 71. It is the same as $y = 2x$ when y is plotted against x.

If the equation is $y = 2x^2 + 3$ on plotting in the same way, that is y against x^2, the resulting line will not pass through the origin but will have an intercept of 3 units on the y axis. It will be the same as $y = 2x + 3$, when y is plotted against x.

The same procedure will be followed with any other power of x. Thus in $y = 2x^3 + 5$, y is plotted against x^3.

In general for the function $y = ax^n + b$ plot y against x^n.

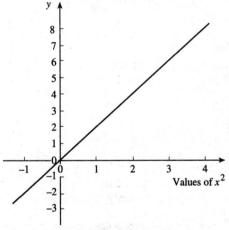

Fig. 71

If $x^n = z$ then the equation takes the form of

$$y = az + b.$$

This is the equation of a straight line. The graph, while not showing the usual relation between y and x as graphs usually do, will make it possible to find the values of a and b by methods previously given.

191 Worked example

Two variables, x and y, are thought to be connected by a law of the form of $y = ax^2 + b$. The following values of x and y are known. Find the law connecting the variable.

x	0.5	1	1.5	2	2.5
y	−9.25	−7	−3.25	2	8.75

As y must be plotted against x^2 you calculate the following table of corresponding values of x^2 and y.

x^2	0.25	1	2.25	4	6.25
y	−9.25	−7	−3.25	2	8.75

Plotting y against x^2, the resulting graph is as shown in Fig. 72.

This is a straight line, and the values of a and b can be found by the method of Section 178.

By inspection of the graph, the intercept on the y-axis (i.e., b) is −10, and a, the gradient of the line, is 3.

∴ the law is $\qquad y = 3x^2 - 10.$

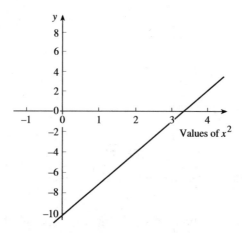

Fig. 72

192 $y = ax^n$. USE OF LOGARITHMS

As was pointed out, the previous method can be used only when the power of x involved is known. If, however, the power is not known, and the law is of the form $y = cx^n$, this can be reduced to the form of the equation of a straight line by taking logarithms.

If $$y = cx^n$$
then $$\log y = n \log x + \log c.$$

Comparing this with the standard form of the equation of a straight line, namely, $y = ax + b$, it is seen to be linear and of the same form, $\log y$ taking the place of y, and $\log x$ taking the place of x.

The constants to be determined are now n and $\log c$, so you plot the graph of

$$\log y = n \log x + \log c.$$

From this graph you can find n and $\log c$ in the same way as a and b in the standard form. When $\log c$ is known you find c from a calculator and the law can be written down.

It is possible to deal with only the simpler cases in this book.

193 WORKED EXAMPLE

Two variables, x and y, are connected by a law of the form $y = cx^n$. The following table gives corresponding values of x and y. Find the law connecting x and y.

x	18	20	22	24	25
y	623	863	1160	1519	1724

Since the law connecting these is of the form

$$y = cx^n$$
taking logs $$\log y = n \log x + \log c.$$

Tabulating values of $\log x$ and $\log y$ in base 10, you get the following:

$\log x$	1.255	1.301	1.342	1.380	1.398
$\log y$	2.795	2.936	3.065	3.182	3.236

The logs correspond in order to the numbers in the column

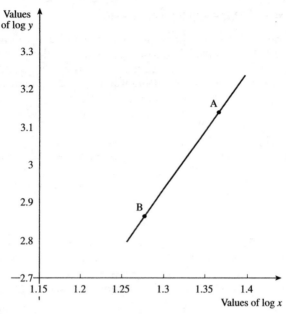

Fig. 73

above, and are calculated approximately to 3 places of decimals.
The graph is the straight line shown in Fig. 73.

Selecting the points A and B on the straight line, you substitute
their coordinates in turn in the equation

$$\log y = n \log x + \log c.$$

Thus you get the equations

$$3.236 = 1.398n + \log c$$
$$2.795 = 1.255n + \log c.$$

Subtracting, $0.441 = 0.143n.$

$$\therefore \qquad n = \frac{0.441}{0.143}$$

$$= 3.1 \text{ approx.}$$

Substituting for n in equation above

$$2.795 = (3.1 \times 1.255) + \log c.$$

$$\therefore \qquad \log c = 2.795 - 3.891$$
$$= -1.0955.$$

$\therefore c = 0.08$ approximately.
$\therefore$ the law connecting y and x is

$$y = 0.08x^{3.1}.$$

EXERCISE 64

1   The variables x and y are connected by a law of the form
    $y = ax^2 + b$. The following corresponding values of x and y
    are known. Find the law.

x	0.5	1	1.5	2	2.5
y	4.5	9	16.5	27	40.5

2   The following table gives related values of x and y.
    Determine whether these values are connected by an equa-
    tion of the form $y = ax^2 + b$ and, if so, find the values of a
    and b.

x	4	5	6	7	8	9
y	14.3	18	22.5	28	34.5	41.5

3   The following values of R and V are possibly connected by a
    law of the type $R = aV^2 + b$. Test if this is so and find the
    law.

V	12	16	20	22	24	26
R	6.44	7.56	9	9.84	10.76	11.76

4   H is connected with V by an equation of the form
    $H = aV^3 + b$. The following corresponding values are
    known. Find the values of a and b.

V	10	12	14	15	17
H	1500	2300	3400	4000	5700

5   The following corresponding values of x and y were mea-
    sured. There may be errors of observation. Test if there is a
    probable law $y = a + bx^2$ and, if this is the case, find the
    probable values of a and b.

x	1	1.5	2	2.30	2.50	2.70	2.80
y	0.77	1.05	1.50	1.77	2.03	2.25	2.42

6 In measuring the resistance, R ohms, of a carbon-filament lamp at various voltages, V, the following results were obtained:

V	60	70	80	90	100	120
R	70	67.2	65	63.3	62	60

The law is thought to be of the form $R = \dfrac{a}{V} + b$. Test this and find a and b.

7 The values of x and y in the following table are connected by a law of the form $y = ax^n$. Find a and n and so determine the law.

x	2	3	4	5
y	2	6.75	16	31.25

8 The following table gives corresponding values of two variables x and y. The law which connects them is of the form $y = ax^n$. Find this law.

x	2	3	5	6	8	10
y	4.24	5.20	6.71	7.35	8.49	9.49

9 The following values of H and Q are connected by a law of the type $Q = aH^n$. Find a and n.

H	1.2	1.6	2.0	2.2	2.5	3
Q	6.087	6.751	7.316	7.571	7.927	8.467

10 Two quantities x and y are connected by an equation of the form $y = ax^n$. The following table gives corresponding values of the variables. Determine a and n.

x	3	3.5	4	4.5	5
y	6.19	6.79	7.35	7.89	8.40

23

Rational and Irrational Numbers.

Surds

194 RATIONAL AND IRRATIONAL NUMBERS

A number which cannot be expressed as an integer, or a ratio of two (finite digit) integers is called an **irrational number**.

As an example of an irrational number, consider $\sqrt{2}$. It can be expressed as a decimal number to any required degree of accuracy, these *approximations* being rational numbers.

Thus to 4 significant figures $\sqrt{2} \approx 1.414 = 1414/1000$.

$\qquad\qquad 6 \qquad\qquad\qquad \sqrt{2} \approx 1.414\,21 = 141421/100\,000$.

$\qquad\qquad 12 \qquad\qquad\qquad \sqrt{2} \approx 1.414\,2135 = 14142135/10\,000\,000$.

It is clear that the greater the required accuracy, the larger the integers have to be, top and bottom, in order to approximate $\sqrt{2}$ with integer fractions. However, it can be proved (by methods beyond the scope of this book) that it is impossible to write $\sqrt{2}$ *exactly* as the ratio of two integers, with a finite number of digits, even though very good *estimates* can be obtained by fractions involving very large integers. Thus $\sqrt{2}$ is irrational.

Note that if a number has a decimal representation which does not contain a finite number of digits, this does not necessarily mean that the number is irrational. A simple example of this is

$$\tfrac{1}{3} = 0.333333\ldots$$

Although the number $\frac{1}{3}$ has a decimal representation which contains 3 recurring indefinitely, it is clearly the ratio of two integers and so is certainly a rational number.

Other roots such as $\sqrt{3}$, $\sqrt{11}$, $\sqrt{19}$, $\sqrt[3]{4}$, ... are examples of irrational numbers. A root of a rational number which is irrational, as these are, is called a **surd**.

There are also other numbers, that do not involve roots, which are irrational. It has been proved that the ratio of the circumference of a circle to its diameter, denoted by the symbol π, cannot be expressed exactly as a ratio of integers with a finite number of figures. It is often expressed roughly by $\frac{22}{7}$, or 3.1416 to 5 significant figures. Computers can evaluate π as a decimal to an immense number of places, but it is impossible to write down an exact value.

However, enough decimal places of π are known so that for all practical calculations this is not actually a problem!

Such numbers as this are irrational, but not surds. They are also called incommensurable.

Since they do not possess fractional representations with finite integers, the infinite number of decimal digits of an irrational number can only be represented approximately by computers. This can sometimes be the origin of mistakes in very sensitive numerical calculations using computers, but skilled programmers can often find algorithms to avoid this problem.

195 IRRATIONAL NUMBERS AND THE NUMBER LINE

When you use the decimal system with a finite number of decimal places to represent an irrational number, you can never specify the number exactly. You can however, state to any required degree of accuracy the limits between which they lie. Thus $\sqrt{2}$ lies on the line between 1.41 and 1.415.

Or, more accurately, $\sqrt{2}$ lies between 1.4142 and 1.4143.

Or, more accurately still, $\sqrt{2}$ lies between 1.414 21 and 1.414 22.

And so on.

196 GEOMETRICAL REPRESENTATION OF SURDS

It may be noted, however, that it is theoretically possible in many cases to obtain by geometrical constructions, straight lines which do represent surds accurately.

For example, from geometry you know that the square on the hypotenuse of a right-angled triangle is equal to the sum of the squares on the sides containing the right angle. Consequently if you construct a right-angled triangle, the sides of which are of unit length, the length of the hypotenuse is $\sqrt{1^2+1^2}$ – i.e., $\sqrt{2}$.

Similarly the hypotenuse of a right-angled triangle of sides 1, and $\sqrt{2}$ units will be $\sqrt{3}$ units. In this way it is possible to represent many surds by straight lines. Theoretically you can mark the lengths thus obtained on the number scale, but in practice no high degree of accuracy can be obtained in the construction of them.

197 Operations with surds

It is the custom in algebra to classify such a number as $\sqrt{a}$ as a surd, though until a numerical value has been assigned to a we cannot say whether or not it is irrational. For purposes of operation, however, it is treated as a surd.

In operating with surds one principle is fundamental. Surds must obey the laws of algebra as formulated for rational numbers. Since you can write surds as powers with fractional indices, e.g. $\sqrt{2} = 2^{\frac{1}{2}}$, you can operate with surds as with these powers, according to the laws of indices.

For example, just as $(a + b)^{\frac{1}{2}}$ is not equal to $a^{\frac{1}{2}} + b^{\frac{1}{2}}$, so $\sqrt{a + b}$ is not equal to $\sqrt{a} + \sqrt{b}$. In this respect the root sign has the same effect as a bracket; the expression under it must be regarded as a whole.

(1) Multiplication

$$\sqrt{a} \times \sqrt{b} = a^{\frac{1}{2}} \times b^{\frac{1}{2}} = \sqrt{a \times b}.$$

Thus:

$$\sqrt{3} \times \sqrt{7} = \sqrt{3 \times 7} = \sqrt{21}$$
$$\sqrt{2}(\sqrt{5} - \sqrt{3}) = (\sqrt{2} \times \sqrt{5}) - (\sqrt{2} \times \sqrt{3}) = \sqrt{10} - \sqrt{6}$$
$$4\sqrt{7} = \sqrt{16} \times \sqrt{7} = \sqrt{16 \times 7} = \sqrt{112}$$
$$3x^2\sqrt{y} = \sqrt{9x^4} \times \sqrt{y} = \sqrt{9x^4y}$$
$$(\sqrt{a} + \sqrt{b})^2 = (\sqrt{a})^2 + 2(\sqrt{a} \times \sqrt{b}) + (\sqrt{b})^2 = a + b + 2\sqrt{ab}$$
$$(\sqrt{a} - \sqrt{b})^2 = a + b - 2\sqrt{ab}$$
$$(\sqrt{a} + \sqrt{b})(\sqrt{a} - \sqrt{b}) = (\sqrt{a})^2 - (\sqrt{b})^2 = a - b$$
$$(\sqrt{5} + 7)(\sqrt{3} - 2) = (\sqrt{5} \times \sqrt{3}) - 2\sqrt{5} + 7\sqrt{3} - 14.$$

By using the above and the converse rules we obtain useful transformations in operations.

Thus:

$$\sqrt{1000} = \sqrt{100 \times 10} = \sqrt{100} \times \sqrt{10} = 10\sqrt{10}$$
$$\sqrt{72} = \sqrt{36 \times 2} = \sqrt{36} \times \sqrt{2} = 6\sqrt{2}$$
$$\sqrt{9a^3b^2} = \sqrt{9 \times a^2 \times a \times b^2} = 3ab\sqrt{a}.$$

The above transformations may also be employed to simplify expressions involving surds.

■ *Examples*

(1) $\sqrt{5} + \sqrt{20} = \sqrt{5} + \sqrt{4 \times 5} = \sqrt{5} + 2\sqrt{5} = 3\sqrt{5}.$

(2) $\sqrt{27} - \sqrt{75} + \sqrt{48} = \sqrt{9 \times 3} - \sqrt{25 \times 3} + \sqrt{16 \times 3}$
$$= 3\sqrt{3} - 5\sqrt{3} + 4\sqrt{3}$$
$$= 2\sqrt{3}.$$

(2) **Rationalisation**

The evaluation of a number such as $\dfrac{1}{\sqrt{2}}$ will be easier, if the fraction can be transformed so that you multiply by the surd and do not divide by it. You can do this by the following transformation:

$$\frac{1}{\sqrt{2}} = \frac{\sqrt{2}}{\sqrt{2} \times \sqrt{2}} = \frac{\sqrt{2}}{2}.$$

Similarly

$$\frac{\sqrt{5}}{2\sqrt{3}} = \frac{\sqrt{5} \times \sqrt{3}}{2\sqrt{3} \times \sqrt{3}} = \frac{\sqrt{15}}{2 \times 3} = \frac{\sqrt{15}}{6}.$$

By this transformation the denominator is changed from an irrational number to a rational one.

This is called rationalising the denominator.

If the denominator is a binomial expression the method is slightly more difficult. The procedure is indicated in the following examples:

■ *Example 1:* Rationalise the denominator of $\dfrac{1}{\sqrt{5} - \sqrt{2}}$.

Since $(a - b) \times (a + b) = a^2 - b^2$ and $(\sqrt{a} + \sqrt{b}) \times (\sqrt{a} - \sqrt{b}) = a - b$ (see above), then, if the denominator is multiplied by $\sqrt{5} + \sqrt{2}$, the surds will disappear from it.

Thus
$$\frac{1}{\sqrt{5} - \sqrt{2}} = \frac{\sqrt{5} + \sqrt{2}}{(\sqrt{5} - \sqrt{2})(\sqrt{5} + \sqrt{2})}$$

$$= \frac{\sqrt{5} + \sqrt{2}}{5 - 2} = \frac{\sqrt{5} + \sqrt{2}}{3}.$$

Example 2: Simplify $\dfrac{\sqrt{5}-1}{\sqrt{5}+1}$.

To rationalise the denominator it must be multiplied by $\sqrt{5}-1$.

$$\therefore \quad \frac{\sqrt{5}-1}{\sqrt{5}+1} = \frac{\sqrt{5}-1)(\sqrt{5}-1)}{(\sqrt{5}+1)(\sqrt{5}-1)} = \frac{(\sqrt{5}-1)^2}{(\sqrt{5})^2-(1)^2}$$

$$= \frac{5-2\sqrt{5}+1}{5-1}$$

$$= \frac{6-2\sqrt{5}}{4} = \frac{3-\sqrt{5}}{2} .$$

EXERCISE 65

1 Express the following as complete square roots, thus:

$$3\sqrt{7} = \sqrt{9\times7} = \sqrt{63}.$$

    (1) $5\sqrt{6}$ (2) $12\sqrt{2}$
    (3) $10\sqrt{5}$ (4) $4\sqrt{13}$
    (5) $2a\sqrt{b}$ (6) $3x^2\sqrt{y}$

2 Express the following with the smallest possible number under the root sign, in each case:

    (1) $\sqrt{800}$ (2) $\sqrt{320}$
    (3) $\sqrt{108}$ (4) $\sqrt{5000}$
    (5) $\sqrt{375}$ (6) $\sqrt{7200}$
    (7) $\sqrt{24a^3b^2}$ (8) $\sqrt{18x^4y^3z}$
    (9) $\sqrt{75a^3b^3c^3}$ (10) $\sqrt{1000p^2q}$
    (11) $\sqrt{a^2b(a-b)^3}$ (12) $\sqrt{25xy^3(x-2y)^2}$

3 Simplify:

    (1) $\sqrt{3} \times \sqrt{15}$ (2) $\sqrt{14} \times \sqrt{7}$
    (3) $\sqrt{32} \times \sqrt{24}$ (4) $2\sqrt{5} \times 5\sqrt{2}$
    (5) $\sqrt{2}(2\sqrt{2}-1)$ (6) $\sqrt{7}(\sqrt{14}+\sqrt{2})$
    (7) $\sqrt{a+b} \times \sqrt{a^2-b^2}$
    (8) $\sqrt{3(x-2y)} \times \sqrt{6(x^2-4y^2)}$

4 Multiply the following:

    (1) $(\sqrt{2}-1)(\sqrt{2}+3)$
    (2) $(2\sqrt{3}+\sqrt{2})(\sqrt{3}-2\sqrt{2})$

(3) $(\sqrt{2} - 1)^2$ (4) $(2\sqrt{5} + \sqrt{3})^2$

(5) $(\sqrt{7} - 5\sqrt{2})^2$

(6) $(\sqrt{5} + \sqrt{3})(\sqrt{5} - \sqrt{3})$

(7) $(2\sqrt{10} - \sqrt{2})(2\sqrt{10} + \sqrt{2})$

(8) $(1 + 10\sqrt{3})^2$ (9) $(\sqrt{a} - 5)(\sqrt{a} + 5)$

(10) $(\sqrt{27} + \sqrt{6})(\sqrt{27} - \sqrt{6})$

5   Simplify the following by rationalising the denominator:

(1) $\dfrac{2}{\sqrt{3}}$ (2) $\dfrac{1}{\sqrt{5}}$

(3) $\dfrac{3}{2\sqrt{7}}$ (4) $\dfrac{1}{3\sqrt{10}}$

(5) $\dfrac{1}{\sqrt{32}}$ (6) $\dfrac{12}{\sqrt{20}}$

(7) $\dfrac{\sqrt{2}}{\sqrt{3}}$ (8) $\dfrac{2\sqrt{5}}{\sqrt{7}}$

(9) $\dfrac{1}{\sqrt{5} - 1}$ (10) $\dfrac{3}{\sqrt{7} + \sqrt{2}}$

(11) $\dfrac{3}{2\sqrt{3} - \sqrt{2}}$ (12) $\dfrac{5}{2\sqrt{7} - 3\sqrt{3}}$

(13) $\dfrac{\sqrt{3} + \sqrt{2}}{\sqrt{3} - \sqrt{2}}$ (14) $\dfrac{\sqrt{7} - \sqrt{5}}{\sqrt{7} + \sqrt{5}}$

(15) $\dfrac{\sqrt{7}}{3\sqrt{7} - 1}$ (16) $\dfrac{1}{\sqrt{2} - 1} + \dfrac{1}{\sqrt{2} + 1}$

(17) $\dfrac{1}{\sqrt{5} + \sqrt{3}} - \dfrac{1}{\sqrt{5} - \sqrt{3}}$

(18) $\dfrac{4}{6 - \sqrt{5}}$

24

Arithmetical and Geometrical Sequences

198 MEANING OF A SEQUENCE

A sequence is a succession of numbers each of which is formed according to a definite law, which is the same throughout the sequence. In the past, the word series was also used to mean sequence.

The ordinary numbers 1, 2, 3, 4, ... constitute a sequence, each term of which is greater by unity than the one which immediately precedes it.

5, 9, 13, 17, ... is a sequence, each term of which is greater by 4 than the one which immediately precedes it.

2, 4, 8, 16, ... is a sequence in which each term is twice the one which immediately precedes it.

$1, \frac{1}{2}, \frac{1}{3}, \frac{1}{4}, \ldots$ is a sequence in which the terms are the reciprocals of 1, 2, 3, 4, ...

$1^2, 2^2, 3^2, \ldots$

$1^3, 2^3, 3^3, \ldots$

are sequences, the construction of each of which is obvious.

199 THE FORMATION OF A SEQUENCE

Sequences are of great importance in modern mathematics, but in this book it is possible to deal with only a few simple cases.

The two most important things to be known about a sequence usually are:

(1) The law of its formation. If this is known, it is possible to find any term in the sequence.
(2) The sum of a given number of terms of the sequence.

In this connection it is necessary to consider what is the nature of the sum when the number of terms is very great. If the sequence is one in which the terms increase numerically such as

$$2, 5, 8, 11,\ldots$$
or
$$1, 3, 9, 27, \ldots$$

it is clear that the more terms which are taken, the greater will be the sum. But if the sequence is one in which the terms decrease as the number of terms increase, such as $1, \frac{1}{3}, \frac{1}{9}, \frac{1}{27}, \ldots$, it is not always easy to discover what the sum will be when the number of terms is very great. This is a matter which will be considered later.

200 ARITHMETIC SEQUENCES, OR ARITHMETIC PROGRESSIONS

An arithmetic sequence, or arithmetic progression, is one in which each term is formed from that immediately preceding it by adding or subtracting a constant number.

The number thus added or subtracted is called the common difference of the sequence.

■ *Examples:*

(1) $7, 13, 19, 25, \ldots$ (common difference 6).
(2) $6, 4, 2, 0, -2, \ldots$ (common difference -2).

In general if three numbers a, b, c, are in arithmetic progression (denoted by A.P.) then

$$b - a = c - b.$$

201 ANY TERM IN AN ARITHMETIC SEQUENCE

Let $a =$ the first term of an arithmetic sequence
  $d =$ the common difference ($+$ve or $-$ve)
then the sequence can be written

$$a, a + d, a + 2d, a + 3d, \ldots$$

It is evident that the multiple of d which is added to a to produce any term is one less than the number of the term in the sequence.

Thus the fourth term is $a + (4 - 1)d = a + 3d$. Hence if the number of any term is denoted by n

then $\qquad$ nth term $= a + (n - 1)d$.

■ *Examples:*

(1) In the sequence 7, 10, 13, . . . the common difference is 3.
  ∴ the tenth term is $7 + (10 - 1)3 = 34$.
  ∴ the nth term is $\quad 7 + (n - 1)3$.
(2) In the sequence 6, 2, -2, -6, . . . $d = -4$.
  ∴ the nth term $\quad = 6 + (n - 1)(-4) = 6 - (n - 1)4$.
  ∴ the eighth term $= 6 + 7(-4) = 6 - 28 = -22$.

202 THE SUM OF ANY NUMBER OF TERMS OF AN ARITHMETIC SEQUENCE

Using the following symbols in addition to those used previously:

Let n = number of terms whose sum is required
  s = the sum of n terms
  l = the last term

then by the previous formula

$$l = a + (n - 1)d.$$

Now $\quad s = a + (a + d) + (a + 2d) + \ldots + (l - d) + l.$

Reversing the series

$$s = l + (l - d) + (l - 2d) + \ldots + (a + d) + a.$$

Adding the corresponding terms of the two sets, each pair gives $(a + l)$.

$2s = (a + l) + (a + l) + (a + l) + \ldots + (a + l) + (a + l)$
  $= (a + l) \times n$, since there are n terms and ∴ n pairs.

∴ $\qquad$ $$s = \frac{n(a + l)}{2}.$$

Since $l = a + (n - l)d$, on substituting for l in the last result

$$s = \frac{n\{a + a + (n - 1)d\}}{2}.$$

∴ $\qquad$ $$s = \frac{n}{2}\{2a + (n - 1)d\}.$$

This formula, like all other formulae, may be used not only to find s, but also any of the other numbers n, a, or d.

To find a and d offers no difficulty, but if n is required it will be seen that a quadratic equation will result. Since there are two roots to every quadratic equation, two values of n will always be found. In some cases only one root is admissible; in others both roots provide solutions.

For example, in a sequence involving negative terms such as

$$9 + 7 + 5 + 3 + 1 - 1 - 3$$

you can see that the sum of 7 terms is the same as the sum of 3 terms. In other cases it will be evident that one of the roots is inadmissible.

203 ARITHMETIC MEAN

If three numbers are in arithmetic progression, the middle one is called the **arithmetic mean** of the other two.

Let a, b, c, be three numbers in A.P.

Then by the definition of Section 200

$$b - a = c - b$$
$$2b = a + c.$$
$$\therefore \qquad b = \frac{a + c}{2} .$$

It will be seen that the arithmetic mean of two numbers is the same as their average.

It is usual also to speak of inserting arithmetic means between two numbers, by which is meant that they, together with the two given, form a series of numbers in A.P.

■ *Example:* Insert three arithmetic means between 4 and 20.

If these are a, b, c, then 4, a, b, c, 20 are in A.P., five terms in all.
   Using $l = a + (n - 1)d$ for the fifth term 20
      $20 = 4 + (5 - 1)d$, whence $d = 4$.

∴ the five terms are 4, 8, 12, 16, 20.

204 WORKED EXAMPLES

■ *Example 1:* The sum of an A.P. of 8 terms is 90 and the first term is 6. What is the common difference?

Using $$s = \frac{n}{2}\{2a + (n - 1)d\}.$$

and substituting given values

$$90 = \tfrac{8}{2}\{(2 \times 6) + (8 - 1)d\}$$
$$90 = 4(12 + 7d) = 48 + 28d.$$
$$\therefore \qquad 28d = 42$$
and $$d = 1.5.$$

■ *Example 2:* How many terms of the series 3, 6, 9, ... must be taken so that their sum is 135?

Using $$s = \frac{n}{2}\{2a + (n - 1)d\}.$$

and substituting, $$135 = \frac{n}{2}\{6 + (n - 1)3\}.$$

$$\therefore \qquad 270 = n(3 + 3n)$$
or $$270 = 3n + 3n^2.$$
$$\therefore \qquad n^2 + n - 90 = 0.$$

Factorising $(n - 9)(n + 10) = 0$
$$\therefore \qquad n = 9 \text{ or } -10.$$

The root -10 is inadmissible as having no meaning in this connection.

$$\therefore \text{ the solution is} \qquad n = 9.$$

EXERCISE 66

1 Write down the next three terms of the following series:

  (1) $5, 7.5, 10, \ldots$
  (2) $12, 8, 4, \ldots$
  (3) $(a + 3b), (a + b), (a - b)$
  (4) $2.7, 4, 5.3$
  (5) $x - y, x, x + y$

2 Find the fifth and eighth terms of the series whose first term is 6, and common difference 1.5.

3 Find the $2p$th term of the series whose first term is 6 and common difference 2.

4 Find the nth term of the series whose first term is $(x + 2)$ and common difference 3.

5 Find the twenty-fifth term of the series, 0.6, 0.72, 0.84, ...

6 The fourth term of an A.P. is 11 and the sixth term 17. Find the tenth term.

7 The fifth term of an A.P. is 11 and the ninth term is 7. Find the fourteenth term.

8 Which term of the series 2.3, 4.2, 6.1, ... is 36.5?

9 Find the sums of the following series:

(*a*) 15, 16.5, 18, ... to ten terms.
(*b*) 9, 7, 5, ... to eight terms.
(*c*) 0.8, 0.6, 0.4, ... to nine terms.
(*d*) $2\frac{2}{3}, 3\frac{1}{2}, 4\frac{1}{3}, \ldots$ to twenty-seven terms.

10 How many terms of the series 10, 12, 14, ... must be taken so that the sum of the series is 252?

11 How many terms of the series 24, 20, 16, ... must be taken so that the sum of the series is 80?

12 Find the thirtieth term and the sum of thirty terms of the series 4, 8, 12, ...

13 A contractor agrees to sink a well 250 m deep at a cost of £27 for the first metre, £28.50 for the second metre, and an extra £1.50 over the price of the previous metre for each additional metre. Find the cost of the last metre and the total cost.

14 A girl's parents place in the savings bank £25 on her first birthday, £50 on her second, £75 on her third, and so on, increasing the amount by £25 on each birthday. How much will be saved up when the girl reaches her sixteenth birthday, the latter inclusive?

205 HARMONIC SEQUENCES, OR HARMONIC PROGRESSIONS

A sequence of numbers is said to be in harmonic progression (H.P.) if their reciprocals form a sequence in arithmetic progression.

Thus the sequence   1, 3, 5, 7, ... are in A.P.
∴ the sequence $1, \frac{1}{3}, \frac{1}{5}, \frac{1}{7}, \ldots$ are in H.P.

This sequence is important in the theory of sound.

It is not possible to obtain a simple formula for the sum of n terms of an H.P., but many problems relating to such a sequence can be solved by using the corresponding arithmetic sequence.

Harmonic mean

The harmonic mean of two numbers may be found as follows.

Let a and b be the numbers.
Let H be their harmonic mean,

i.e., a, H, b are in H.P.

Then $\dfrac{1}{a}$, $\dfrac{1}{H}$, $\dfrac{1}{b}$ are in A.P.

$$\therefore \qquad \frac{1}{H} = \frac{1}{2}\left(\frac{1}{a} + \frac{1}{b}\right) \text{(Section 203)}$$

$$= \frac{a+b}{2ab} \ .$$

$$\therefore \qquad H = \frac{2ab}{a+b} \ .$$

206 GEOMETRIC SEQUENCES OR GEOMETRIC PROGRESSIONS

A geometric sequence is one in which the ratio of any term to that which immediately precedes it is constant for the whole sequence.

This ratio is called the common ratio of the sequence. It may be positive or negative. Thus each term of the sequence can be obtained by multiplying the term which precedes it by the common ratio.

■ *Examples*

(1) $1, 2, 4, 8, \ldots$ (common ratio 2).
(2) $1, \frac{1}{2}, \frac{1}{4}, \frac{1}{8}, \ldots$ (common ratio $\frac{1}{2}$).
(3) $2, -6, 18, -54, \ldots$ (common ratio -3).
(4) $R, R^2, R^3, R^4, \ldots$ (common ratio R).

If three numbers a, b, c are in geometric progression (G.P.) then $\dfrac{b}{a} = \dfrac{c}{b}$.

This is the test to apply in order to find if numbers are in G.P.

General form of a geometric sequence

Let a = 1st term.
Let r = common ratio.

Then the sequence is $a, ar, ar^2, ar^3, \ldots$

207 CONNECTION BETWEEN A GEOMETRIC SEQUENCE AND AN ARITHMETIC SEQUENCE

In the geometric sequence

$$a, ar, ar^2, ar^3, \ldots$$

take logs of each term. You then get the sequence:

$$\log a, \log a + \log r, \log a + 2 \log r, \log a + 3 \log r, \ldots$$

This is an arithmetic sequence in which the first term is $\log a$, and the common difference is $\log r$.

∴ the logarithms of the terms of a G.P. form a sequence in A.P.

208 GENERAL TERM OF A GEOMETRIC SEQUENCE

Examining the sequence $a, ar, ar^2, ar^3, \ldots$ you can see that each term of the sequence is the product of a and a power of r the index of which is one less than the number of the term.

∴ if n = any term then the nth term = ar^{n-1}.

If r is negative, $n - 1$ being alternately odd and even, the terms will be alternately negative and positive, assuming a to be positive.

When $n - 1$ is even, n is odd, and the nth-term is +ve.

When $n - 1$ is odd, n is even, and the nth-term is −ve.

Worked examples

■ *Example 1:* Find the seventh term of the series

$$3, 6, 12, \ldots$$

In this series $r = 2$, so using the formula

$$n\text{th term} = ar^{n-1}$$
$$\text{the seventh term} = ar^6 = 3 \times 2^6 = 3 \times 64$$
$$= 192.$$

■ *Example 2:* Find the eighth term of the series

$$2, -6, 18, -54, + \ldots$$

For this series $r = -3$.
Using ar^{n-1},

$$\text{eight term} = 2 \times (-3)^{8-1} = 2 \times (-3)^7$$
$$= -4374.$$

■ *Example 3:* Find the fifth term of the series in which the first term is 100 and the common ratio 0.63.

Using ar^{n-1}, if x = the fifth term

$$x = 100 \times (0.63)^4 = 15.75.$$

■ *Example 4:* The third term of a G.P. is 4.5 and the ninth is 16.2. Find the common ratio.

Using ar^{n-1},

$$\text{third term} = ar^2 = 4.5$$
$$\text{ninth term} = ar^8 = 16.2.$$

Dividing $\quad ar^8 \div ar^2 = 16.2 \div 4.5.$

∴ $\qquad\qquad r^6 = 16.2 \div 4.5.$

∴ $\qquad\qquad r = 1.238.$

209 GEOMETRIC MEAN

If three numbers are in G.P., the middle term is called the **geometric mean** of the other two.

Let a, b, c be three numbers in G.P.

Then by the definition of Section 206

$$\frac{b}{a} = \frac{c}{b}.$$

∴ $\qquad\qquad b^2 = ac,$

and $\qquad\qquad b = \sqrt{ac}.$

EXERCISE 67

1 Write down the next three terms of each of the following sequences:

   (*a*)   4, 10, 25
   (*b*)   16, 4, 1
   (*c*)   16, −24, 36
   (*d*)   0.3, 0.03, 0.003
   (*e*)   3, 0.45, 0.0675

2 Find the seventh term of the sequence 5, 10, 20, ...
3 Find the seventh term of the sequence 6, −4, $2\frac{2}{3}$, ...
4 Find the fifth term of the sequence 1.1, 1.21, 1.331, ...

5 Find the sixth term of the sequence $-0.5, 0.15, -0.045, \ldots$

6 Write down the $2n$th and the $(2n + 1)$th terms of the sequence:

    (1) $a, ar, ar^2, \ldots$
    (2) $a, -ar, ar^2, \ldots$

7 The first term of a G.P. is 1.05 and the sixth term is 1.3401. Find the common ratio.

8 The fifth term of a G.P. is 1.2166 and the seventh term is 1.3159. Find the common ratio.

9 Find the geometric mean in each of the following cases:

    (1)   3 and 5
    (2)   4.2 and 3.6

10 Insert two geometric means between 5 and 13.72.

11 A person receives a salary of £10,000 a year with an increase each year of 5 per cent of the salary for the previous year. How much salary does the person receive during the fifth year?

12 The expenses of a company are £200,000 a year. It is decided that each year they will be reduced by 5 per cent of those for the preceding year. What will be the expenses during the fourth year, the first reduction taking place during the first year?

13 In a geometric sequence the first term is unity and the fifth term is 1.170 approx. Find the common ratio.

14 Insert three terms in geometric progression between 5 and 80.

210 THE SUM OF n TERMS OF A GEOMETRIC SEQUENCE

In addition to the symbols employed above,

let S_n represent the sum of n terms of a G.P.

Then $\quad S_n = a + ar + ar^2 + \ldots + ar^{n-2} + ar^{n-1}$ (1)

    Multiply both sides by r,

then $\quad rS_n = ar + ar^2 + ar^3 \ldots + ar^{n-1} + ar^n$ (2)

    Subtracting (1) from (2),

$$rS_n - S_n = ar^n - a$$
or $\qquad\qquad S_n(r - 1) = a(r^n - 1).$

$$\therefore \qquad S_n = \frac{a(r^n - 1)}{r - 1} \qquad\qquad \text{(A)}$$

If (2) is subtracted from (1) above, the formula becomes:

$$S_n = \frac{a(1 - r^n)}{1 - r} \qquad\qquad \text{(B)}$$

If $r > 1$ and positive, form (A) should be used.
If $r < 1$ or negative, form (B) should be used.

211 WORKED EXAMPLES

■ *Example 1:* Find the sum of seven terms of the sequence

$$2, 3, 4.5, \ldots$$
$$r = \tfrac{3}{2} = 1.5.$$

Using
$$S = \frac{a(r^n - 1)}{r - 1},$$

and substituting
$$S = \frac{2(1.5^7 - 1)}{1.5 - 1} = 64.34.$$

■ *Example 2:* Find the sum of seven terms of the sequence

$$4, -8, 16, \ldots$$
$$r = -2.$$

Using
$$S = \frac{a(1 - r^n)}{1 - r},$$

and substituting
$$S = \frac{4\{1 - (-2)^7\}}{1 - (-2)} = \frac{4(1 + 128)}{3}$$
$$= \tfrac{4}{3} \times 129 = 172.$$

EXERCISE 68

1   Find the sums of the following sequences

    (*a*)   1.5, 3, 6, ... to six terms.
    (*b*)   30, -15, $7\tfrac{1}{2}$, ... to eight terms.
    (*c*) $\tfrac{1}{2}$, $-\tfrac{1}{4}$, $\tfrac{1}{8}$, ... to six terms.

2 Find the sum of the first six terms of the sequence

$$5, 2.5, 1.25, \ldots$$

3 Find the sum of the first six terms of the sequence

$$1 - \tfrac{2}{3} + \tfrac{4}{9} + \ldots$$

4 Find the sum of the first six terms of the sequence

$$1 + 1.4 + 1.96 + \ldots$$

5 Find the sum of the first twelve terms of the sequence

$$4 + 5 + 6.25 + \ldots$$

6 If the first and third terms of a G.P. are 3 and 12, find the sum of eight terms.

7 If the third and fourth terms of a G.P. are $\tfrac{1}{9}$ and $\tfrac{1}{3}$, respectively, find the eighth term and the sum of eight terms.

8 Find the sum of $20 + 18 + 16.2 + \ldots$ to six terms.

212 INCREASING GEOMETRIC SEQUENCES

When the common ratio of a geometric sequence is numerically greater than unity, as in the sequences

$$1, 2, 4, 8, \ldots$$
$$2.5, 7.5, 22.5, \ldots$$

the terms increase in magnitude. The sum of n terms increases as n increases. If the number of terms increases without limit – that is, n is greater than any number you may select, however great – then the sum of these terms will also increase without limit, i.e. it will become infinitely great, or, to use the mathematical term, approach 'infinity', which is denoted by the symbol ∞.

You can then say that, as n, the number of terms, approaches infinity, S_n, the sum of these terms, also approaches infinity. This may be expressed by the following notation.

If $\qquad\qquad n \to \infty$, then $S_n \to \infty$.

213 DECREASING GEOMETRIC SEQUENCES

If, however, the common ratio is numerically less than unity, as in the following sequences,

$$\tfrac{1}{3}, \tfrac{1}{6}, \tfrac{1}{12}, \ldots$$
$$0.2, 0.02, 0.002, \ldots$$

then as the number of terms increases, the terms themselves decrease. Using the terms employed above, you can say that, as n increases without limit, the terms themselves decrease without limit, and ultimately become indefinitely small.

You cannot say, however, that the sum of these terms increases without limit, as n increases without limit. That is a matter for further investigation.

214 RECURRING DECIMALS

There is an example, arising from arithmetic, which will assist in coming to conclusions on this important question, namely that of a recurring decimal. You know that

$$\tfrac{1}{9} = 0.1111\ldots$$

in which 1 recurs without limit.

The decimal is in effect the sum of a geometric sequence

$$\frac{1}{10} + \frac{1}{10^2} + \frac{1}{10^3} + \ldots$$

in which there is no limit to the number of terms. It is an example of what is called an infinite sequence. But you know that the sum of all these terms, no matter how many are taken, is ultimately equal to the finite fraction $\tfrac{1}{9}$.

If you find the sum of finite numbers of terms, you get:

$$S_1 = \frac{1}{10}\ .$$

$$S_2 = \frac{1}{10} + \frac{1}{10^2} = \frac{11}{10^2}\ .$$

$$S_3 = \frac{1}{10} + \frac{1}{10^2} + \frac{1}{10^3} = \frac{111}{10^3} \text{ and so on.}$$

The difference between $\tfrac{1}{9}$ and the sum of these is:

$$\tfrac{1}{9} - S_1 = \tfrac{1}{90}$$
$$\tfrac{1}{9} - S_2 = \tfrac{1}{900}$$
$$\tfrac{1}{9} - S_3 = \tfrac{1}{9000}$$

and in general, finding the sum of n terms by using the formula

$$S_n = \frac{a(1 - r^n)}{1 - r}$$

you get $\quad S_n = \dfrac{\dfrac{1}{10}\left(1 - \dfrac{1}{10^n}\right)}{1 - \dfrac{1}{10}}$

$$= \frac{1}{9}\left(1 - \frac{1}{10^n}\right) = \frac{1}{9} - \frac{1}{9 \times 10^n}.$$

Examining these results, you can see that the difference between $\frac{1}{9}$ and the various sums, $S_1, S_2, S_3, \ldots S_n$ decreases as n increases. In general, the difference between $\frac{1}{9}$ and the sum of n

terms is $\dfrac{1}{9 \times 10^n}$.

As n increases without limit, this difference becomes closer and closer to zero – and the sum approaches to equality with $\frac{1}{9}$. It can never be greater than $\frac{1}{9}$.

Using the previous notation, you can express this result in the form

$$\text{as } n \to \infty, S_n \to \tfrac{1}{9}.$$

There is thus a limit to which S_n approaches and which it cannot exceed.

215 A GEOMETRICAL ILLUSTRATION

The approach of the sum of a geometric sequence to a limit may be illustrated by a graphical representation of the sequence

$$\frac{1}{2} + \frac{1}{4} + \frac{1}{8} + \frac{1}{16} + \ldots$$

or $\qquad \dfrac{1}{2} + \dfrac{1}{2^2} + \dfrac{1}{2^3} + \dfrac{1}{2^4} + \ldots$

Let the rectangle ABCD (Fig. 74) represent a unit of area.
Let E be the midpoint of AB and draw EF perpendicular to DC.
Then rectangle AEFD represents $\frac{1}{2}$ of a unit.

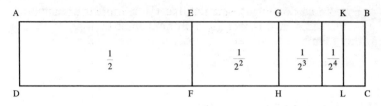

Fig. 74

Bisecting the rectangle EBCF by GH, then rectangle EGHF represents $\frac{1}{4}$ or $\frac{1}{2^2}$ of a unit.

Continuing the process of bisecting the rectangle left over after each bisection, you get a series of rectangles whose areas represent the terms of the above series. These rectangles diminish in area as you represent more and more terms of the series in this way.

The rectangle AKLD represents the sum of the four terms of the series

$$\frac{1}{2} + \frac{1}{2^2} + \frac{1}{2^3} + \frac{1}{2^4} \, .$$

As more divisions are made and the sum of more terms represented by a rectangle, this rectangle approaches nearer to the area of the whole rectangle – i.e., 1 – but can never exceed it. Consequently 1 is a limit which the sum of the series approaches as the number of terms is increased without limit, but which it can never exceed, no matter how many terms are taken.

If you sum the sequence $\frac{1}{2} + \frac{1}{2^2} + \frac{1}{2^3} + \dots$ by using the formula

$$S_n = \frac{a(1 - r^n)}{1 - r} \, ,$$

you get $\qquad S_n = \frac{\frac{1}{2}\{(1 - (\frac{1}{2})^n\}}{1 - \frac{1}{2}} = \frac{\frac{1}{2}(1 - (\frac{1}{2})^n\}}{\frac{1}{2}} \, ,$

$\therefore \qquad S_n = 1 - (\tfrac{1}{2})^n.$

Examining this result, you see that $(\frac{1}{2})^n$ decreases in magnitude as n increases. If n is increased without limit then $(\frac{1}{2})^n$ decreases without limit – i.e., it becomes closer and closer to zero.

$\therefore$ you can say that $S_n \to 1$ as $n \to \infty$.

216 THE SUM TO INFINITY

The above suggests the general treatment of this question.

Using
$$S_n = \frac{a(1 - r^n)}{1 - r}$$

i.e.,
$$S_n = \frac{a - ar^n}{1 - r}$$

you have
$$S_n = \frac{a}{1 - r} - a.\frac{r^n}{1 - r} \,.$$

Considering the term $a.\dfrac{r^n}{1 - r}$, if r lies between $+1$ and -1 then r^n diminishes as n increases or, with the previous notation,
$$\text{as } n \to \infty, \; r^n \to 0 \text{ and } a.\frac{r^n}{1 - r} \to 0.$$

Thus the right hand side approaches $\dfrac{a}{1 - r}$ as a limit.

This is the 'limiting sum' of the series and it is called the 'sum to infinity'.

If it is represented by S_∞

then
$$S_\infty = \frac{a}{1 - r} \,.$$

217 WORKED EXAMPLES

■ *Example 1:* Sum to infinity the series
$$2 + \tfrac{1}{2} + \tfrac{1}{8} + \ldots$$

Here $a = 2$, $r = \frac{1}{4}$.

$\therefore$
$$S_\infty = \frac{2}{1 - \frac{1}{4}} = 2 \div \tfrac{3}{4}$$
$$= 2\tfrac{2}{3}.$$

■ *Example 2:* Find the sum to infinity of the series

$$5 - 1 + \tfrac{1}{5} - \dots$$

Here $a = 5, r = -\tfrac{1}{5}$.

$$\therefore \qquad S_\infty = \frac{5}{1 - (-\tfrac{1}{5})} = 5 \div 1\tfrac{1}{5}$$

$$= 4\tfrac{1}{6}.$$

EXERCISE 69

1 To what limits will the sums of the following sequences tend as the number of terms increases indefinitely?

    (a) $\tfrac{1}{2} + \tfrac{1}{8} + \tfrac{1}{32} + \dots$

    (b) $\tfrac{1}{4} + \tfrac{1}{16} + \tfrac{1}{64} + \dots$

    (c) $\tfrac{1}{2} + \tfrac{1}{4} + \tfrac{1}{8} + \tfrac{1}{16} + \tfrac{1}{32} + \dots$

2 To what limits will the sums of the following sequences tend as the number of terms increases indefinitely?

    (a) $0.1 + 0.001 + 0.000\,01 + \dots$

    (b) $0.06 + 0.006 + 0.000\,06 + \dots$

    (c) $0.16 + 0.0016 + 0.000\,016 + \dots$

What is the connection between these sequences?

3 To what limit does an infinite number of terms in the following sum tend?

$$1 - \tfrac{1}{2} + \tfrac{1}{4} - \tfrac{1}{8} + \dots$$

Show the connection with the sequence in question 1.

4 Show that the sum of n terms of the sequence

$$1 + \tfrac{1}{3} + \tfrac{1}{9} + \dots$$

is $\tfrac{3}{2}\{1 - (\tfrac{1}{3})^n\}$.

Hence show what limit this sequence approaches.

5 Find the sum of n terms of the sequence

$$1 + \frac{a}{b} + \frac{a^2}{b^2} + \dots$$

when

$$\frac{a}{b} < 1.$$

Hence find the limit approached by the sum as the number of terms becomes infinitely great.

6 Find the limiting sum, or the sum to infinity of the following sequences:

   (a) $\frac{8}{3} + \frac{4}{9} + \frac{2}{27} + \ldots$
   (b) $5 - 1 + \frac{1}{5} - \ldots$
   (c) $9 - 6 + 4 - \ldots$

7 Find the sum to infinity of the sequence

   (a) $1 + \dfrac{1}{1.04} + \dfrac{1}{1.04^2} + \ldots$

   (b) $\sqrt{2} + 1 + \dfrac{1}{\sqrt{2}} + \ldots$

8 The sum to infinity of a sequence is 15, and the first term is 3. Find the common ratio.

9 A superball is dropped from a height of 10 m. At each rebound it rises to a height which is 0.9 of the height from which it has just fallen. What is the total distance through which the ball will have moved before it finally comes to rest?

10 The yearly output of a silver mine is found to be decreasing by 25 per cent of its previous year's output. If in a certain year its output was £25,000,000 what could be reckoned as its total future output?

218 SIMPLE AND COMPOUND INTEREST

The accumulation of money when put to interest furnishes examples of arithmetic and geometric sequences. When money is put out at simple interest, the interest is payable for each year, but is not added to the principal.

For example the interest payable on £100 at 5 per cent for 1, 2, 3 … years will be £5 each year. The total interest that has been paid after 1, 2, 3 … years will be £5, £10, £15, these sums forming a sequence in A.P. The total interest paid varies directly as the time.

But if the money is lent at compound interest, the interest is added each year to the principal, and for the following year the interest is calculated on their sum.

Suppose £1 is invested at 5 per cent compound interest. Then the interest for the first year is £$\frac{5}{100}$ or £0.05.

∴ the amount at the end of the year is £1.05 and the amount at the end of the year of £P is £$P \times 1.05$.

Consequently the ratio of the amount at the end of a year to that at the beginning is always 1.05.

This corresponds to the ratio of a geometric sequence.

∴ amount at the end of the second year is

$$P \times 1.05 \times 1.05 = P \times 1.05^2.$$

∴ amount at the end of the third year is

$$P \times 1.05^2 \times 1.05 = P \times 1.05^3.$$

∴ amount at the end of the fourth year is

$$P \times 1.05^3 \times 1.05 = P \times 1.05^4.$$

∴ amount at the end of the nth year

$$= P \times 1.05^n.$$

These amounts at the end of successive years, namely:

$$P \times 1.05, P \times 1.05^2, P \times 1.05^3 \ldots$$

constitute a geometric sequence.

Let M = the amount at the end of n years

then $$M = PR^n.$$

In this formula, as you have seen in other cases, any one of the four quantities may be the subject of the formula.

Thus $$P = \frac{M}{R^n}.$$

This enables you to find the principle amount of money which will produce £M in n years.

Again $$R^n = \frac{M}{P}.$$

∴ $$R = \sqrt[n]{\frac{M}{P}},$$

from which you can find the rate of interest.

Again $$n \log R = \log M - \log P.$$

∴ $$n = \frac{\log M - \log P}{\log R}$$

which enables you to find the time taken for P to amount to M.

219 ACCUMULATED VALUE OF PERIODICAL PAYMENTS

Suppose that £P is invested each year for 10 years at 5 per cent C.I., each investment being made at the beginning of a year.

Using the above formula:

The first £P at the end of 10 years amounts to

$$P \times 1.05^{10}.$$

The second £P at the end of 9 years amounts to

$$P \times 1.05^{9}.$$

The third £P at the end of 8 years amounts to

$$P \times 1.05^{8}.$$

And finally, the last £P invested bears interest for 1 year and amounts to

$$P \times 1.05.$$

Then the accumulated value of the investments amounts to

$$P \times 1.05^{10} + P \times 1.05^{9} + P \times 1.05^{8} \ldots + P \times 1.05,$$

or, reversing the sequence,

$$P \times 1.05 + P \times 1.05^{2} + P \times 1.05^{3} \ldots + P \times 1.05^{10}.$$

This is the sum of a geometric sequence and using the form

$$S_n = \frac{a(r^n - 1)}{r - 1}$$

the accumulated value of the investments is

$$\frac{1.05(1.05^{10} - 1)}{1.05 - 1} \times P.$$

220 ANNUITIES

An annuity is a sequence of equal annual payments extending over a specified number of years, or for the life of the annuitant.

A ground rent is a similar financial transaction, the holder of the freehold receiving an annual payment, called ground rent, for the number of years specified in the lease. Ground rents and annuities are constantly being bought and sold, and the method of calculating the amount to be paid by the purchaser can be determined by means of the above results. This amount will depend upon the rate of interest which the purchaser expects to receive on his investment.

Suppose that the rate of interest expected is 4 per cent.

The price is obtained by finding the present value of each of the payments, as follows:

From the formula $M = PR^n$

you get $P = \dfrac{M}{R^n}$.

P is the amount which produces M in n years at the given rate per cent; it is called the present value of M due in n years.

If £A is the annual payment and P is its present value, then for the first payment due in 1 year

$$P = \frac{A}{R} = \frac{A}{1.04}$$

when the rate per cent is 4.

For the second payment

$$P = \frac{A}{1.04^2}.$$

For the third payment

$$P = \frac{A}{1.04^3},$$

and so on.

∴ the total present value

$$= \frac{A}{1.04} + \frac{A}{1.04^2} + \frac{A}{1.04^3} + \dots$$

$$= A\left\{ \frac{1}{1.04} + \frac{1}{1.04^2} + \frac{1}{1.04^3} + \dots \right\}$$

$$= A\left\{ \frac{\dfrac{1}{1.04}\left[1 - \left(\dfrac{1}{1.04}\right)^n\right]}{1 - \dfrac{1}{1.04}} \right\}$$

$$= \frac{A}{0.04}\left\{ 1 - \frac{1}{1.04^n} \right\}$$

This can then be evaluated for any value of A.

The terms of the above sequence decrease and if the ground rent is a perpetual one, or the lease is a very long one, the present value becomes the sum to infinity of the above sequence – i.e.,

Present value

$$= A \left\{ \frac{\dfrac{1}{1.04}}{1 - \dfrac{1}{1.04}} \right\}$$

$$= A \left\{ \frac{\dfrac{1}{1.04}}{\dfrac{1.04 - 1}{1.04}} \right\} = A \frac{1}{0.04} = A \times \frac{100}{4} = A \times 25.$$

The ground rent is then said to be worth 25 years purchase. It is always found by dividing 100 by the rate per cent.

EXERCISE 70

1 If £1,000 is invested at the beginning of each year for 10 years at 3 per cent C.I., find the accumulated value a year after the last amount is invested.

2 An annuity of £6,000 a year is allowed to accumulate at 3 per cent C.I. for 8 years. What was the total amount at the end?

3 A man saves £250 every half-year and invests it at C.I. at $4\frac{1}{2}$ per cent. What will be the amount of his savings in 8 years if the last amount saved bears interest for 6 months?

4 Find the present value of an annuity of £3,000 for 10 years, reckoning C.I. at 4 per cent, the first payment being due one year after purchase.

5 What should be the purchase price of an annuity of £5,000 for 8 years, reckoning C.I. at $3\frac{1}{2}$ per cent?

6 A woman wished to endow in perpetuity an institution with a yearly sum of £2,000. If C.I. is reckoned at 4 per cent, what amount will be needed for it?

7 A man retires at 65, when the expectation of life is 10.34 years, with a pension of £20,000. What single payment would be the equivalent of this, reckoning C.I. at 4 per cent?

8 A pension of £6,000 per annum was awarded to Nelson and his heirs for ever. If this is commuted into a single payment, what should that be, reckoning C.I. at $2\frac{1}{2}$ per cent?

Appendix

The following brief statement of permutations, combinations and the binomial theorem, and a note on the roots of a quadratic equation are given for the benefit of students who may need to use them in the differential calculus or other branches of more advanced mathematics.

PERMUTATIONS AND COMBINATIONS

1 Permutations

Consider the following example:

A party of 6 people arrived at a theatre and obtained 4 seats together and 2 separate. In how many different ways could the 4 seats in a row be filled if there are no restrictions as to where any of the 6 may sit?

Consider the first seat. Since any one of the 6 people may sit in it, it can be filled in 6 different ways. With each of these 6 ways, the second seat can be filled in 5 different ways, since 5 people are left.

∴ there are (6×5) different ways of filling the first two seats.

With each of the 6×5 or 30 ways of filling the first two seats, there are 4 ways of filling the third seat, since 4 people are left.

∴ there are $(6 \times 5 \times 4)$ different ways of filling the first three seats.

Similarly the fourth seat can be filled in 3 ways and

∴ there are $(6 \times 5 \times 4 \times 3)$ different ways of filling the 4 seats – i.e., 360 ways.

Arrangements of a number of different objects in a row are called **permutations** and the above problem was that of the permutation of 6 things 4 at a time.

This is expressed by a special notation – namely, $^6P_4$ or $_6P_4$.

It will be seen that if the people mentioned had 6 seats together, the number of permutations or arrangements in these seats would be $6 \times 5 \times 4 \times 3 \times 2 \times 1$. This product of all the integral numbers from 1 to 6 inclusive is called factorial 6 and is expressed by 6!

In general, the product of the integral numbers from 1 to n inclusive is denoted by $n!$

Thus $n! = n(n - 1)(n - 2) \ldots \times 3, \times 2, \times 1$ and is called factorial n.

2 Permutations of *n* things *r* at a time, or ^nP_r

This is the general treatment of the above special case and the method adopted to find the formula is the same.

There are r places to be filled and n different things to choose from.

The 1st place can be filled in n ways
The 2nd place can then be filled in $(n - 1)$ ways

since with each of the n ways of filling the first place, each of the $(n - 1)$ ways of filling the second can be associated.

$\therefore$ there are $n(n - 1)$ ways of filling the first two places.

Similarly there are

$n(n - 1)(n - 2)$ ways of filling the three places,
and $n(n - 1)(n - 2)(n - 3)$ ways of filling the four places.

$\therefore$ by inspection there are

$n(n - 1)(n - 2)(n - 3) \ldots \{n - (r - 1)\}$ ways of filling the r places.

$\therefore {}^nP_r = n(n - 1)(n - 2)(n - 3) \ldots (n - r + 1)$.

If the n things are all arranged among themselves, then the last factor becomes $(n - n + 1)$ or 1.

$\therefore \qquad\qquad\qquad {}^nP_n = n!$

3 Combinations

The problem solved above – namely, the number of different ways of filling up 4 seats by 6 people – might have been approached in another way.

(1) You could find the number of different sets or groups of 4 that could be formed from 6 people.
(2) Each group could then be arranged in the seats in 4! ways.

The product of these two numbers must give the total number of ways of filling the 4 seats – i.e., the permutations of 6 things 4 at a time. The difficulty at present is that of finding the number of groups.

Let x = the number of groups

then, by the above reasoning,

$$x \times 4! = {}^6P_4.$$
$$\therefore \qquad x = {}^6P_4 \div 4!$$

Thus you can find the number of groups, when you know the number of permutations.

Such groups are called combinations.

A number of groups or combinations of 6 different things 4 at a time is denoted by $\binom{6}{4}$, and in general, the number of combinations of n things r at a time is denoted by $\binom{n}{r}$. The notation $\binom{n}{r}$ is read as 'n choose r'. Older notations used are ${}^nC_r$ and ${}_nC_r$.

4 The Combinations of *n* things *r* at a time

With the same reasoning as that employed above you can deduce that

$$\binom{n}{r} = {}^nP_r \div r!$$

or

$$\binom{n}{r} = \frac{n(n-1)(n-2)\ldots(n-r+1)}{r!}.$$

Thus

$$\binom{n}{2} = \frac{n(n-1)}{2!}$$

$$\binom{n}{3} = \frac{n(n-1)(n-2)}{3!}$$

$$\binom{10}{4} = \frac{10 \times 9 \times 8 \times 7}{1 \times 2 \times 3 \times 4} = 210$$

$$\binom{8}{3} = \frac{8 \times 7 \times 6}{1 \times 2 \times 3} = 56.$$

THE BINOMIAL THEOREM

5 Products of binomial factors

It was shown in Section 91 that

$$(x + a)(x + b) = x^2 + x(a + b) + ab.$$

Employing the methods used in Chapter 13 we can show that
$(x + a)(x + b)(x + c)$
$= x^3 + x^2(a + b + c) + x(ab + bc + ca) + abc.$

You should note that:

(1) the expression is arranged in descending powers of x;
(2) the coefficients of these powers, after the first, are the sums formed in every way from using the letters a, b, c,

    (*a*) one at a time, (*b*) 2 at a time, (*c*) 3 at a time.

From the way in which this product is formed you can deduce the product of

$$(x + a)(x + b)(x + c)(x + d).$$

Arranging the powers of x in descending order, the coefficients of these powers will be

x^4 – unity.
x^3 – sum of the letters one at a time, i.e.,

$$(a + b + c + d).$$

x^2 – sum of the letters two at a time, i.e.,

$$(ab + bc + ad + bd + cd + ac).$$

x – sum of the letters three at a time, i.e.,

$$(abc + bcd + acd + abd).$$

The term independent of x is $abcd$.

$\therefore$ the full product is
$x^4 + x^3(a + b + c + d) + x^2(ab + bc + cd + ad + bd + ac) +$
$$x(abc + bcd + acd + abd) + abcd.$$

You saw in Section 4, that the number of ways of

(1) Grouping 4 letters 1 at a time is $\binom{4}{1} = \dfrac{4}{1} = 4.$

(2) Grouping 4 letters 2 at a time is $\binom{4}{2} = \dfrac{4.3}{1.2} = 6.$

(3) Grouping 4 letters 3 at a time is $\binom{4}{3} = \dfrac{4.3.2}{1.2.3} = 4$.

(4) Grouping 4 letters 4 at a time is $\binom{4}{4} = 1$.

In the above factors let $b = c = d = a$
then the left side is $(x + a)^4$.

In the expansion of it:

The coefficient of x^3 is $(a + a + a + a) = 4a$.
The coefficient of x^2 is $6a^2$.
The coefficient of x is $4a^3$.

The last term is a^4.

$\therefore \quad (x + a)^4 = x^4 + 4x^3a + 6x^2a^2 + 4xa^3 + a^4$

or $\quad (x + a)^4 = x^4 + \binom{4}{1}x^3a + \binom{4}{2}x^2a^2 + \binom{4}{3}xa^3 + a^4$.

By a similar process you can obtain the expansion

$$(x + a)^5 = x^5 + \binom{5}{1}x^4a + \binom{5}{2}x^3a^2 + \binom{5}{3}x^2a^3 + \binom{5}{4}xa^4 + a^5.$$

From a consideration of these results you can deduce the general case – namely,

$$(x + a)^n = x^n + \binom{n}{1}x^{n-1}a + \binom{n}{2}x^{n-2}a^2 + \binom{n}{3}x^{n-3}a^3 + \ldots + a^n$$

or

$$(x + a)^n = x^n + nx^{n-1}a + \frac{n(n - 1)}{2!}x^{n-2}a^2$$

$$+ \frac{n(n - 1)(n - 2)}{3!}x^{n-3}a^3 + \ldots + a^n.$$

This is called the binomial theorem.

The above reasoning is independent of the values of x and a. It will therefore hold if a is replaced by $-a$. Then

$$(x - a)^n = x^n + nx^{n-1}(-a) + \frac{n(n - 1)}{1.2}x^{n-2}(-a)^2 + \ldots + (-a)^n.$$

Since odd powers of $(-a)$ are negative and even powers are positive, the terms will be alternately +ve and −ve if $a > 0$.

$$\therefore (x - a)^n = x^n - nx^{n-1}a + \frac{n(n - 1)}{1.2}. x^{n-2}a^2$$

$$- \frac{n(n - 1)(n - 2)}{1.2.3}x^{n-3}a^3 + \ldots + (-a)^n.$$

In the previous results let $x = 1$.

Then

$$(1 + a)^n = 1 + na + \frac{n(n - 1)}{1.2}a^2 + \frac{n(n-1)(n - 2)}{1.2.3}a^3 + \ldots + a^n$$

and

$$(1 - a)^n = 1 - na + \frac{n(n - 1)}{1.2}a^2 - \frac{n(n-1)(n - 2)}{1.2.3}a^3 + \ldots + (-a)^n.$$

Every binomial expression can be reduced to one of these forms as follows:

$$(x + a)^n = \left\{x\left(1 + \frac{a}{x}\right)\right\}^n = x^n\left(1 + \frac{a}{x}\right)^n$$

$$= x^n\left\{1 + n.\frac{a}{x} + \frac{n(n - 1)}{1.2}\left(\frac{a}{x}\right)^2 + \ldots + \left(\frac{a}{x}\right)^n\right\}$$

Similarly,

$$(x - a)^n = x^n\left(1 - \frac{a}{x}\right)^n.$$

A complete proof of the binomial theorem requires a more advanced knowledge of algebra than is provided in this book. The demonstration given above assumes that n is a positive integer. In using the theorem later the important question will arise:

Is the binomial theorem true for fractional and negative indices?

It can be proved that the form holds for all values of n. For example of $n = \frac{1}{2}$.

$$(1 + x)^{\frac{1}{2}} = 1 + \frac{1}{2}x + \frac{\frac{1}{2}(\frac{1}{2} - 1)}{1.2}x^2 + \frac{\frac{1}{2}(\frac{1}{2} - 1)(\frac{1}{2} - 2)}{1.2.3}x^3 + \ldots$$

When n is a positive integer, a term in the sequence will ultimately be reached when one of the factors

$$n(n - 1)(n - 2) \ldots$$

will become $(n - n)$ and will vanish, as will all succeeding products.

The number of terms therefore is finite and will clearly be $n + 1$.

Consequently when a value is assigned to x in $(1 + x)^n$ the sum of the sequence will be a finite number and the sequence is said to be convergent.

But if n is fractional or negative, none of the factors

$n(n-1)(n-2)$ ... will vanish, and the number of terms is infinite. If the sum increases without limit as n increases the sequence is said to be divergent. But it can be shown that if x, in $(1+x)^n$, is such that, the sum of the sequence will approach a limit, as is the case with certain geometric sequences (see Section 216). Subject to this condition the sequence $-1 < x < 1$ is always said to be convergent.

THE ROOTS OF A QUADRATIC EQUATION

Writing the general quadratic equation

$$ax^2 + bx + c = 0$$

in the form
$$x^2 + \frac{b}{a}x + \frac{c}{a} = 0 \qquad \text{(A)}$$

the solution is $x = \dfrac{-b \pm \sqrt{b^2 - 4ac}}{2a}$.   (Section 143)

Let α and β be the two roots

then $\alpha + \beta = \dfrac{-b + \sqrt{b^2 - 4ac}}{2a} + \dfrac{-b - \sqrt{b^2 - 4ac}}{2a}$

$$= -\frac{2b}{2a} = -\frac{b}{a} .$$

= coeff. of x in equation (A).

Again $\alpha\beta = \dfrac{-b + \sqrt{b^2 - 4ac}}{2a} \times \dfrac{-b - \sqrt{b^2 - 4ac}}{2a}$

$$= \frac{(-b)^2 - (\sqrt{b^2 - 4ac})^2}{4a^2} \qquad \text{(Section 97)}$$

$$= \frac{4ac}{4a^2}$$

= term independent of x in (A).

Summarising, $\alpha + \beta = -\dfrac{b}{a}$

$$\alpha\beta = \frac{c}{a}$$

Nature of the roots of a quadratic equation

If $(b^2 - 4ac)$ is negative, the square root has no arithmetical meaning. It is customary to speak of such a root as imaginary, while the square root of a positive number is called real.

Hence, for the roots of a quadratic,

(1) If $b^2 > 4ac$, the roots are real and different.
(2) If $b^2 = 4ac$ the roots are equal.
(3) If $b^2 < 4ac$ the roots are imaginary.

A number such as $\sqrt{-p}$ where $p \geq 0$ can be written

$$\sqrt{p \times (-1)} = \sqrt{p} \times \sqrt{-1}.$$

The number $\sqrt{-1}$ is usually denoted by i.
Hence, as an example, $\sqrt{-9}$ can be written $\pm\sqrt{9 \times -1}$

$$= \pm 3i.$$

■ *Example:* The roots of the equation $x^2 + 2x + 5 = 0$ are given by

$$x = \frac{-2 \pm \sqrt{4 - 20}}{2}$$

$$= \frac{-2 \pm \sqrt{-16}}{2}$$

$$= \frac{-2 \pm 4i}{2}$$

$$= -1 \pm 2i.$$

Note: It will be found that the graph of $y = x^2 + 2x + 5$ does not cut the x axis, i.e., it does not equal zero for any real value of x.

Answers

Exercise 1 (p. 11)

1 (1) $100x$ (2) $\dfrac{n}{100}$

2 $\dfrac{100a}{n}$ **3** $100an + mb$

4 (1) $1000a$ (2) $2x$ (3) $\dfrac{y}{1000}$

5 $28 - n$ **6** $x + 50$, or $50 - x$ **7** $\dfrac{a}{x}$

8 (1) x m   (2) $\dfrac{180\,000}{x}$

9 $2n + 5$ **10** (1) $x + 2, x - 2$ (2) $x + 1, x - 1$

11 $\dfrac{xa + yb}{100}$ **12** $\dfrac{2x + 5}{6y}$

13 $a - b$ **14** xy

15 $a + b$ **16** $\dfrac{x + 2}{y - 5}$

17 mv km $+ nu$ km   **18** $100a + b$

19 xv km; $\dfrac{y}{v}$ hours   **20** The original numbers are the digits of the final result, m and n.

Exercise 2 (p. 18)

1 (*a*) 10 doz.; 120   (*b*) $10a$; 120
2 (*a*) $5 \times 73 = 365$ (*b*) $5b = 365$
3 $3a + 4b$; 47   **4** $19a$; 47.5
5 (1) $26b$ (2) $19x$ (3) $12a$ (4) $4x$
6 (1) $2a + 4b$ (2) $12p + 3q$ (3) $6a + 3b + 4$

7 (1) $10a + 2b$ (2) $2b + c - d$ (3) $8x + 2z$
8 (1) 13   (2) 7   (3) 13   (4) 11
9 (1) $8ab = 64$ (2) $3ax + bx = 42$
   (3) $3xy = 45$ (4) $4ab + 4bx - ay = 36$
10 $3\frac{3}{4}$ **11** 22
12 (1) 66   (2) 29   (3) 30
13 $n + 2, n + 4, n + 6; 3n + 12$ **14** $a, 2a, 4a, 8a; 15a$
15 $a, a + d, a + 2d, a + 3d, a + 4d; 5a + 10d$
16 $2n + 5, 2n + 8, 2n + 11, 2n + 14, 2n + 17; 10n + 55$

Exercise 3 (p. 22)

1 $12a$ **2** $10xy$ **3** $2xy$

4 $21mn$ **5** $\dfrac{ab}{12}$ **6** $4ab$

7 $60abc$ **8** $\dfrac{xyz}{24}$ **9** x^3

10 a^4 **11** x^4 **12** a^6
13 $2a^5$ **14** $6x^7$ **15** $2a^2b^2$
16 $6b^9$ **17** x^3y^3 **18** $7x^6a^2$
19 $6a^6$ **20** $27a^6b^3$ **21** x^6
22 $8a^{12}$ **23** $16a^{12}$ **24** $64a^4$
25 54   **26** 21   **27** 8
28 172   **29** 33   **30** 1
31 216   **32** 576

Exercise 4 (p. 24)

1 a^2 **2** a^2 **3** $3x$ **4** $\dfrac{b^4}{2}$

5 $2a^4$ **6** $5y^3$ **7** $3x^3$ **8** $2c^2$
9 ab **10** x^3y **11** $5a^2b$ **12** $3xy$
13 2   **14** $\frac{5}{4}$

Exercise 5 (p. 27)

1 $\dfrac{12a}{35}$ **2** $\dfrac{x}{15}$

3 $\dfrac{4y + 5x}{xy}$ **4** $\dfrac{19a}{6b}$

5 $\dfrac{61x}{40x}$

6 $\dfrac{31x}{18y}$

7 $\dfrac{20a^3 + 9b^2c}{24a^2bc^2}$

8 $\dfrac{3x^2 + 4}{3x}$

9 $\dfrac{6ab^2 + 5}{2b}$

10 $\dfrac{40y^2 - 6}{5y^2}$

11 $\dfrac{a^2 + a + 1}{a}$

12 $\dfrac{x^2 - x + 1}{x}$

13 $\dfrac{c + 2a + 3b}{abc}$

14 $\dfrac{5bc - 4ac + 2ab}{a^2b^2c^2}$

15 x^2y

16 $\dfrac{4b}{a^2}$

17 $\dfrac{3b^2}{2a^2c}$

18 1

19 xy

20 $\dfrac{a}{bc}$

21 $\dfrac{4}{3a}$

22 $\dfrac{3y}{2}$

23 $\dfrac{9}{5q}$

24 $\dfrac{2xy}{3}$

25 $\dfrac{2x^2}{y}$

26 $2y$

27 1

28 $\dfrac{x^2}{y^2}$

29 $\frac{2}{3}$

30 $\dfrac{2x^4}{3y^2}$

31 $\dfrac{x^2}{y^2}$

32 $\dfrac{2a^3}{3b^2}$

33 $\dfrac{8a^2}{3c}$

34 $\dfrac{3y^2}{4}$

35 $\dfrac{20}{xy}$

36 $\dfrac{4a^2b^2}{c^4}$

Exercise 6 (p. 33)

1 $15x + 18z$

2 $6a^2 + 8ab$

3 $18a^3 + 42a^2b - 36a^2c$

4 $8x + y$

5 $5x^3 + 4x^2$

6 x

7 $8x + 5y - 4z$

8 $x - 2y - z$

9 $2x - y + 2z$

10 $a + 7b$

11 $a - b$

12 $a + b$

13 $4x + 2y - 2z$

14 $a + 5b - 5c$

15 $3y$

16 $a^2 + b^2$

17 $x^3 + x^2y - x^3y + xy^3$ **18** $x^2 + 9x + 23$
19 $9p^2 + 3pq + 15q^2$ **20** $5x^2y^2 - 3xy + 6x^2$
21 $2x^4 + 8x^3$

Exercise 7 (p. 35)

1 $6a - 9$ **2** $3a + 30b$ **3** $\frac{3}{2}$
4 $3a^2 + 2ab - 2ac$ **5** $2p^2 + p$
6 $x^2 + 3xy + 12y^2$ **7** $7bc - 2b^2 + 2c^2$
8 $13x + 5$ **9** $34 - 6a + 4b$ **10** $x + y$

11 $2a - 10b$ **12** $\dfrac{3x^2}{2}$ **13** $\dfrac{27c}{8}$

14 $2a - (b - c)$ **15** $x - (y + z)$ **16** $2(a + 2b - 3c)$
17 $x^2 - y(x - y)$ **18** $2a - 4b + 7c$ **19** $x + 2y - 2z$
20 (1) 111   (2) 6

Exercise 8 (p. 43)

1 $0 + 4 - 2 + 4 - 6 = 0$
2 $+8 + 2 - 14 + 4 - 6 = -6°C$
3 (1) 11°C   (2) 15°C   (3) 30°C
4 (1) +5   (2) −5   (3) +8
5 (1) (*a*) +2   (*b*) +4   (2) +5
6 (1) (+4)   (2) (+8)   (3) (−8)
     (4) (−4)   (5) (+3)   (6) (−8)
     (7) 0   (8) (−8)
7 (1) $7a$ (2) $-7x$ (3) $10ab$
     (4) $-x - 8y$ (5) $a - 7b$ (6) $-x + 2y$
     (7) $7x - 3y$ (8) -8
8 (1) $2x - 2y$ (2) $2x - y - 7z$
9 (1) $3a - (-5a)$ (2) $5x - (6x)$
     (3) $-3a - (-10a)$
10 (1) $(-a)$ (2) $(+a)$ (3) $(-a)$

Exercise 9 (p. 47)

1 (1) (+36)   (2) (−36)   (3) (−36)
     (4) (+36)   (5) (+4)   (6) (−4)
     (7) (−4)   (8) (+4)
2 (1) $(-a^2)$ (2) $(+a^2)$ (3) (−1)
     (4) (+1)
3 (1) $(-4ab)$ (2) $(+4ab)$ (3) $(-20xy)$

(4) $(-20xy)$ (5) $\left(-\dfrac{5x}{y}\right)$ (6) $\left(+\dfrac{5x}{y}\right)$

4 (1) $+24$ (2) $+6a^3$
(3) $+3y$ (4) $-6ab$
5 (1) $-10x^3$ (2) $a^2 - b^2$
(3) $-2ab^2$ (4) $-a^2 + 2ab + ac$
6 (1) $+20ab^2$ (2) $6x^2 + 8x$
(3) $-2xz + 2yz$
7 (1) $a^2, -a^3, a^4, -a^5$
(2) $4x^2, -8x^3, +16x^4, -32x^5$

(3) $\dfrac{b^2}{9}, -\dfrac{b^3}{27}, \dfrac{b^4}{81}, -\dfrac{b^5}{243}$

8 $\pm 9, \pm 3x^2, -x, -2a^2$
9 (1) $+16x$ (2) $5x$ (3) $2y$
(4) $-2b$ (5) $-2t$ (6) $2x$
(7) $8x^3$ (8) $-3x$ (9) $-4ab$
(10) $6a^2c$

10 (1) $-b$ (2) $-x$ (3) $-\dfrac{9x}{4}$

Exercise 10 (p. 51)

1 (*a*) 17 (*b*) 47 (*c*) 20 (*d*) 47
2 (*a*) $3n$ (*b*) $3n - 1$ (*c*) $3n - 2$ (*d*) n^2
(*e*) $n^2 + 10$ (*f*) $5n + 1$

Exercise 11 (p. 52)

1 (*a*) 1, 5, 9, 13, 17 (*b*) 16, 20, 24, 28, 32
(*c*) 4, 16, 36, 64, 100
2 (*a*) $-3, 1, 5, 9, 13$ (*b*) 32, 52, 72, 92, 112
(*c*) 32, 44, 64, 92, 128
3 (*a*) 18, 23, 28, 33, 38 (*b*) 11, 23, 35, 47, 59
(*c*) 19, 34, 55, 82, 115
4 (*a*) 17, 24, 31, 38, 45
(*b*) 17, 24, 31, 38, 45; yes; $7(n + 1) + 3 = 7n + 7 + 3 = 7n + 10$

Exercise 12 (p. 53)

1 (*a*) $5(4n - 3)$ (*b*) $5(n + 3) - 2$ (*c*) $2(6n - 5) - 7$
(*d*) $2(3(n + 4) - 5)$
2 (*a*) 11 (*b*) 59 (*c*) 93
3 (*a*) 23 (*b*) 23

Exercise 13 (p. 56)

1   (a) $x \to 2x + 7; 13$ (b) $x \to 2(x + 7); 20$
     (c) $x \to (x + 5)^2; 64$ (d) $x \to x^2 + 5; 14$
2   (a) 2   (b) 18
     (c) $x \to 2x - 3$ (d) $x \to 2(x - 3)$
3   (a) 40   (b) 47   (c) 361   (d) 6566
     (e) $x \to (2x + 7)^2 + 5$

Exercise 14 (p. 57)

1   (a) 23   (b) $x \to (x - 3) \div 4$
2   (a) 23   (b) $x \to (x - 3) \div 5 + 4$
3    $x \to (x + 3) \div 2; 7$
4    $x \to ((x - 7) \div 5 + 3) \div 2; 3$

Exercise 15 (p. 59)

1   (a) 4   (b) 10   (c) 25
2   (a) 6   (b) 5
3   (a) 5   (b) 3

Exercise 16 (p. 65)

1   (a) $2\frac{1}{2}$ (b) 20
2   (a) 16   (b) 60
3   (a) -16 (b) -10
4   (a) 80   (b) -72
5   (a) 12.6   (b) -0.08
6   (a) $3\frac{1}{5}$ (b) 5
7   (a) 2.5   (b) 12
8   (a) 8   (b) $\frac{4}{3}$
9   (a) $\frac{8}{5}$ (b) 16
10   (a) 1.8   (b) -45
11   (a) 14.8   (b) $18\frac{3}{4}$
12   (a) $\frac{10}{9}$ (b) $\frac{1}{3}$
13   22   14   5   15   8   16   29
17   $-\frac{1}{2}$ 18 -3 19 $18\frac{4}{5}$
20   (a) $\frac{32}{5}$ (b) $27\frac{1}{2}$
21   (a) -80 (b) -10
22   -35 23   8   24 -7 25 $\frac{7}{3}$
26   3   27 $\frac{7}{3}$ 28 $-\frac{55}{2}$ 29   2.7 to 1
                                                      decimal place
30   2.9 to 1   31   3   32 $\frac{45}{19}$ 33   36
     decimal place
34   $16\frac{1}{2}$

Exercise 17 (p. 68)

1	12	**2**	72.5	**3**	42
4	15	**5**	13.5	**6**	21, 23, 25
7	10 km		**8**	40	
9	11.9 cm, 10.1 cm		**10**	4 people, 60p	
11	5		**12**	Son 8 years, father 32 years	
13	30°C		**14**	80 at £18 and 40 at £12	
15	23				

Exercise 18 (p. 71)

1 175 **2** 154 **3** $113\frac{1}{7}$

4 85.6 to 1 decimal place **5** 0.616 to 3 decimal places **6** $57\frac{5}{7}$ metres

7 $4\frac{1}{2}$ **8** 204 **9** $\frac{5}{8}$

10 1390 (approx.) to 0 decimal place **11** 31.4 to 1 decimal place

Exercise 19 (p. 76)

1 $r = \sqrt{\dfrac{A}{\pi}}$ **2** $r = 3\sqrt{\dfrac{3V}{4\pi}}$

3 $r = \sqrt{\dfrac{3V}{\pi h}}$ **4** $C = \dfrac{825H}{E}$

5 $B = \sqrt{\dfrac{F \times 112 \times 10^5}{A}}$ **6** $l = 3\sqrt{\dfrac{48EId}{W}}$

7 $s = \dfrac{v^2 - u^2}{2a}$, $17\frac{1}{2}$

8 (1) $V = IR$ (2) $R = \dfrac{V}{I}$, 0.1

9 $n = \sqrt{\dfrac{NR - 1}{r}}$, $\frac{70}{9}$ **10** $k = 1008; v = \dfrac{1008}{p}$

Exercise 20 (p. 78)

1 $\dfrac{4a}{5}$ **2** $2a$ **3** p

4 $4b$ **5** $\dfrac{3a - 2b}{a}$ **6** $\dfrac{bp + c}{b}$

7　$\dfrac{2(b-a)}{b}$　　　　**8**　$\dfrac{6-19b}{3a}$　　　　**9**　p

10　$\dfrac{2b-12}{3a}$　　　　**11**　$\dfrac{3b}{a-b}$　　　　**12**　$\dfrac{12a}{4+3a}$

13　$\dfrac{a^2+b^2}{a-b}$　　　　**14**　$\dfrac{m}{n-3}$

Exercise 21 (p. 84)

1	$x=3, y=6$	**2**	$x=5, y=8$
3	$x=12, y=3$	**4**	$x=8, y=3$
5	$x=3, y=1$	**6**	$x=4, y=5$
7	$x=7, y=4$	**8**	$x=10, y=7$
9	$x=6, y=5$	**10**	$x=3, y=10$
11	$x=6, y=10$	**12**	$a=5, b=-\frac{3}{2}$
13	$a=-2, b=3$	**14**	$p=-2, q=4$
15	$x=\frac{5}{2}, y=-2$	**16**	$x=9, y=13$
17	$a=-7, b=\frac{3}{2}$	**18**	$x=6, y=-4$
19	$x=8, y=12$	**20**	$x=1.35, y=2.7$
21	$x=4, y=2$	**22**	$P=1.8, Q=0.32$
23	$x=\frac{15}{2}, y=-3$	**24**	$x=12, y=6$

Exercise 22 (p. 88)

1　$x=12, y=10$　　　　**2**　16, 12

3　20, 7　　　　**4**　10, 6

5　$m=2, b=-3$. Equation is $y=2x-3$; when $x=6, y=9$.

6　$m=20, b=10$

7　$a=0.5, b=0.6; E=6.5$　　　　**8**　$a=-1.36, b=1.38$ to 2 decimal places

9　10 m, 6 m　　　　**10**　£37.50; 20p

11　Tie £9.80, socks £4.80　　　　**12**　$u=10 \text{ ms}^{-1}, a=6 \text{ ms}^{-2}$, 125 m

Exercise 23 (p. 92)

1　(*a*)　$H<190$ cm　　　　　　(*b*)　$P>50$
　　(*c*)　£200 $\leqslant S <$ £300　　(*d*)　3 hours $< T <$ 5 hours

Exercise 24 (p. 95)

1　(*a*)　$x>7$　　　　　　(*b*)　$x<3$
2　(*a*)　$x<8$　　　　　　(*b*)　$x\leqslant 3$

3 (*a*) $x < 8$ (*b*) $x \leqslant 5$
4 (*a*) $x < 6$ (*b*) $x \geqslant 2$
5 (*a*) $x < 3$ (*b*) $x < -15$
6 $x > 4$
7 $x \geqslant 8$
8 $x < -16$

Exercise 25 (p. 97)

1 $0 \leqslant x \leqslant 5$
2 $2 \leqslant x < 5$
3 $3 < x < 7$

Exercise 26 (p. 98)

1 (*a*) $4 < x < 7$ (*b*) $-2 < x < 1$
2 $-3 \leqslant x < 3$

Exercise 27 (p. 107)

2 10.4°C   **3** Males: (1) 33.9   (2) 18.1
                     Females: (1) 36.8   (2) 20.2
4 (1) 44.2 kg   (2) 65 kg   (3) 1.5 kg
5 Accurate deductions simply on the basis of the graphs (which are anyway not smooth) are certain to be unrealistic. However ...
     (1) 1978–9   (2) 126 000   (3) 1978–9
6 25.1 m/s
7 $P = £133$ is wrong and should be £150. £128; £225; £365
8 About 360 kilotherms.
9

x	-3	-2	-1	0	1	2	3	4	5
y	6	2.5	0	-1.5	-2	-1.5	0	2.5	6

3.45; -1.45.

Exercise 28 (p. 116)

1 (1) 2   (2) -2 **2** (1) -1 (2) $\frac{2}{3}$
3 (1) $-\frac{5}{4}$ (2) $\frac{5}{8}$ **4** (1) 3   (2) 2
5 (1) 4   (2) 6   **6** (1) $-\frac{1}{2}$ (2) 1
7 $a = 2$, $b = 3$. Equation is $y = 2x + 3$. Intercept is 3.
8 $x + 2y = 4$

Exercise 29 (p. 118)

1 A, (4, 4);   B, (4.6, 1.2);   C, (−2, 3);
       D, (4, −2);   E, (−1.4, −3.4);   F, (3, 0);
       G, (0, −3).
2 (1.1 approximately, 1.1 approximately)
4 (2, 2)
5 They lie on a straight line parallel to the x-axis.
6 They appear to lie on a straight line which passes through the
       origin.
7 For every point on it the x-coordinate is +3.

Exercise 30 (p. 125)

1 All lines pass through the origin with different slopes.
2 All lines are parallel with slope 45°.
3 All lines are parallel with gradient $\frac{1}{2}$.
4 All lines have the same intercept on the y-axis, namely 2.
5 Intercepts are (x intercept first):   (1) 1.5 and 3
       (2) 4 and −2   (3) 2 and 5   (4) 2.5 and −2
6 (1) $x = 4, y = 3$ (2) $x = 3, y = 1$
       (3) $x = −2, y = 2$
7 $a = 3; −1$
8 $b = 1; 1$
9 $a = 3, b = −2$. Equation is $y = 3x − 2; −2$.
10 $y = 2x + 3$
11 $3y − 2x = 7$. Gradient $= \frac{2}{3}$.
12 $y − x = 2$. Gradient $= 1$.

Exercise 31 (p. 131)

1 (*a*) yes   (*b*) no   (*c*) yes   (*d*) no
2 (2, 5) is in the region; (6, 1) is not in the region.
3 (3, 4) is in the region; (4, 6) is not in the region.
5 (3, 6) is not in the region; (3, 7) is in the region.
6 (*a*) $y \geqslant 2x + 3, x + 3y \leqslant 12$
       (*b*) $x + 3y \leqslant 12, y \leqslant 2x + 3, x + y \leqslant 6$
       (*c*) $x + 3y \leqslant 12, x + y \geqslant 6$

Exercise 32 (p. 133)

1 (5, 7)
2 (3, 4)

3 $(3, 5)$

4 $(6, 3)$ and $(5, 4)$; $(6, 4)$ is not included since $x + 3y < 18$.

5 $(8, 4)$; $(7, 4)$ is not included since $y < 2x - 10$

Exercise 33 (p. 137)

1 (a) $8L + 3S \geqslant 144$ (b) $L + S \leqslant 36$; £$(48L + 12S)$;
8 large tents and 27 small
tents, £708

2 $L + S \leqslant 100$; $18L + 45S \leqslant 2700$; 67 hectares of lettuces and 33 of spring onions; £10,400; 60 hectares

3 $(4,2), (4,3), (4,4), (5,2), (5,3), (6,2)$

4 (a) $V + L \leqslant 10$ (b) $72V + 180L \leqslant 1080$
(c) $600V + 1100L \geqslant 6600$; $(7,3), (8,2), (9,2), (10,2)$

5 $(16, 2), (14, 3)$ and $(12, 4)$

Exercise 34 (p. 143)

1 $ab + ay + bx + xy$ **2** $ce + cf + de + df$

3 $acxy + adx + bcy + bd$ **4** $ab - ay - bx + xy$

5 $ax - bx - ay + by$ **6** $ab + ay - bx - xy$

7 $ab - ay + bx - xy$ **8** $ab + 3a + 2b + 6$

9 $ab - 3a - 2b + 6$ **10** $ab + 3a - 2b - 6$

11 $ab - 3a + 2b - 6$ **12** $x^2 + 12x + 35$

13 $a^2b^2 + 9ab + 18$ **14** $x^2 + 13x + 30$

15 $x^2 - 13x + 30$ **16** $x^2 + 7x - 30$

17 $x^2 - 7x - 30$ **18** $p^2 - 4p - 96$

19 $x^2 - 12xy + 32y^2$ **20** $x^2 + 4xy - 32y^2$

21 $x^2 - 4xy - 32y^2$ **22** $2a^2 + 9ab + 10b^2$

23 $9x^2 - 27xy + 20y^2$ **24** $28x^2 + 15x + 2$

25 $6x^2 - 11x + 3$ **26** $9x^2 - 9x - 4$

27 $1 - y - 12y^2$ **28** $18x^2 - 27x - 5$

29 $14x^2 + 29xy - 15y^2$ **30** $18a^2 - 57ab + 35b^2$

31 $a + b$ **32** $a + b - c$

33 12   **34** $5y - 5x = 0.5$

35 (a) $x^3 - y^3$ (b) $a^3 + 8$
(c) $1 + x^3$ (d) $x^3 + 3x^2a + 3xa^2 + a^3$

Exercise 35 (p. 146)

1 $x^2 + 4x + 4$ **2** $x^2 - 4x + 4$

3 $a^2 + 6ab + 9b^2$ **4** $a^2 - 6ab + 9b^2$

5 $4x^2 + 4xy + y^2$ **6** $x^2 - 4xy + 4y^2$

7 $a^2b^2 + 20ab + 100$ **8** $x^2y^2 - 6xy + 9$
9 $16x^2 + 40xy + 25y^2$ **10** $16x^2 - 40xy + 25y^2$
11 $25x^2y^2 + 60xy + 36$ **12** $1 - 20x^2 + 100x^4$
13 $25x^4 + 30x^2y^2 + 9y^4$ **14** $9x^2y^2 - 12xy^3 + 4y^4$

15 $x^2 + \dfrac{2x}{y} + \dfrac{1}{y^2}$ **16** $\dfrac{1}{x^2} - \dfrac{2}{xy} + \dfrac{1}{y^2}$

17 $a^2 + \dfrac{2a}{3} + \dfrac{1}{9}$ **18** $\dfrac{1}{4} - \dfrac{3y}{4} + \dfrac{9y^2}{16}$

19 $x^2 + 2xy + y^2 + 2x + 2y + 1$
20 $1 - 2x + 4y + x^2 - 4xy + 4y^2$
21 $a^2 + b^2 + c^2 + 2ab - 2ac - 2bc$
22 $x^2 + y^2 + z^2 - 2xy + 2xz - 2yz$
23 $4x^2 + 9y^2 + 25z^2 + 12xy - 20xz - 30yz$
24 $16a^2 + 4b^2 + 1 - 16ab - 8a + 4b$
25 $x^3 + 3x^2y + 3xy^2 + y^3$ **26** $x^3 - 3x^2y + 3xy^2 - y^3$
27 $a^3 + 6a^2 + 12a + 8$ **28** $a^3 - 6a^2 + 12a - 8$
29 $p^3 + 3p^2q + 3pq^2 + q^3$ **30** $p^3 - 3p^2q + 3pq^2 - q^3$
31 $8x^3 + 12x^2y + 6xy^2 + y^3$ **32** $x^3 - 6x^2y + 12xy^2 - 8y^3$
33 $27a^3 - 27a^2 + 9a - 1$ **34** $1 - 9b + 27b^2 - 27b^3$
35 $2xy$ **36** (1) $4ab$; (2) $-4ab$
37 $40x$ **38** $9y^2 + 18y + 9$; 36
39 $4a(x + a)$; 256 m^2 **40** $2a(x + y + 2a)$

Exercise 36 (p. 147)

1 $a^2 - x^2$ **2** $p^2 - q^2$
3 $a^2 - 4b^2$ **4** $16x^2 - 9$
5 $4x^2 - 1$ **6** $1 - 36x^2$
7 $1 - a^4$ **8** $4x^4 - 1$
9 $x^4 - y^4$ **10** $9x^2y^2 - 4$
11 $144x^2y^2 - 1$ **12** $\frac{1}{16}x^2 - 49$
13 $(x + y)^2 - z^2$ **14** $(a + x)^2 - y^2$
15 $(2a + 3b)^2 - 1$ **16** $(x - 2y)^2 - 36$
17 $a^2 - 4(b + c)^2$ **18** $4x^2 - 9(y + z)^2$

19 $x^2 - \frac{4}{9}$ **20** $\dfrac{x^2}{4} - \dfrac{y^2}{9}$

Exercise 37 (p. 149)

1 $6(x + 2)$ **2** $a(3b + 2)$
3 $2y(2x + y)$ **4** $2a(3a - 2b)$

5 $7xy(2xy - 1)$ **6** $16(1 - 2a^2)$

7 $a(a - b + c)$ **8** $x(x^2 + 3x - 1)$

9 $a^2(15a - 5b + 3b^2)$ **10** $3ac(2a - 5c)$

11 $ab(a + b - c)$ **12** $\dfrac{bc}{3}\left(\dfrac{c}{2} - \dfrac{a}{3}\right)$

13 $7.4(13^2 + a^2)$

14 $18.6(18.6 + 1.4) = 18.6 \times 20 = 372$

Exercise 38 (p. 152)

1 $(a + b)(x + y)$ **2** $(p + q)(c + d)$

3 $(a + d)(b + e)$ **4** $(x - y)(a - c)$

5 $(x + p)(x + q)$ **6** $(x - g)(x - h)$

7 $(a + 5)(b + 6)$ **8** $(a - 5)(b - 6)$

9 $(a - 5)(b + 6)$ **10** $(2a + 3)(b - 5)$

11 $(ax - b)(x + a)$ **12** $(x - b)(x + a)$

Exercise 39 (p. 154)

1 $(x + 2)(x + 1)$ **2** $(x - 2)(x - 1)$

3 $(x + 3)(x + 2)$ **4** $(x - 3)(x - 2)$

5 $(x + 6)(x + 1)$ **6** $(x + 5)(x + 4)$

7 $(x - 10y)(x - 2y)$ **8** $(a - 12b)(a - 3b)$

9 $(xy + 6)(xy + 9)$ **10** $(ab - 16)(ab - 3)$

11 $(y - 9)(y - 12)$ **12** $(x - 7y)(x - 5y)$

13 $(x - 2)(x + 1)$ **14** $(x + 2)(x - 1)$

15 $(x + 3y)(x - 2y)$ **16** $(x - 3y)(x + 2y)$

17 $(b - 3)(b + 1)$ **18** $(b + 3)(b - 1)$

19 $(x + 16)(x - 3)$ **20** $(x - 16)(x + 3)$

21 $(x - 11y)(x + 10y)$ **22** $(a - 12)(a + 1)$

23 $(a - 4)(a + 3)$ **24** $(p + 9)(p - 8)$

25 $(p - 36)(p + 2)$ **26** $(1 - 5x)(1 - 4x)$

27 $(1 - 10x)(1 + 2x)$ **28** $(xy - 11)(xy + 8)$

29 $(p + 9)(p - 5)$ **30** $(p - 7q)(p + 8q)$

Exercise 40 (p. 155)

1 $(3x + 4)(x + 2)$ **2** $(4x - 3)(3x - 2)$

3 $(6x + 1)(2x - 5)$ **4** $(9x - 2)(x + 5)$

5 $(2x + 1)(x + 1)$ **6** $(3x - 1)(x - 1)$

7 $(2x + 1)(x + 2)$ **8** $(3x + 1)(2x + 1)$

9 $(2x - 3)(2x - 1)$ **10** $(5x - 1)(x - 1)$

11 $(2x - 3)(3x - 1)$ **12** $(4x + 1)(3x + 2)$
13 $(2a - 1)(a + 1)$ **14** $(2a + 1)(a - 1)$
15 $(2a + 3)(a - 2)$ **16** $(5b + 2)(2b - 1)$
17 $(5b - 2)(2b + 1)$ **18** $(2y - 5)(4y + 3)$
19 $(4x - 1)(3x + 2)$ **20** $(7c + 2)(2c - 3)$

Exercise 41 (p. 158)

1 $(p + q)^2$ **2** $(x - 2y)^2$
3 $(3x + 1)^2$ **4** $(4x - 5y)^2$

5 $\left(x + \dfrac{1}{2}\right)^2$ **6** $\left(\dfrac{a}{3} + \dfrac{b}{2}\right)^2$

7 $(a + b + 2)^2$ **8** $(x - y - 5)^2$
9 $(x + 10)(x - 10)$ **10** $(ab + 5)(ab - 5)$
11 $(2x + 3y)(2x - 3y)$ **12** $(5a + 4b)(5a - 4b)$
13 $(11x + 6y)(11x - 6y)$ **14** $(12p + 13q)(12p - 13q)$
15 $(5 + 4a)(5 - 4a)$ **16** $(1 + 15x)(1 - 15x)$
17 $2(2a + 5b)(2a - 5b)$ **18** $3(x + 5)(x - 5)$
19 $5(x + 3y)(x - 3y)$ **20** $(a + b + c)(a + b - c)$
21 $(x + 2y + 4z)(x + 2y - 4z)$
22 $(1 + \tfrac{9}{4}y)(1 - \tfrac{9}{4}y)$ **23** $(x + y + z)(x - y - z)$
24 $(a + x - 2y)(a - x + 2y)$ **25** $(x - 15)(x - 1)$
26 $4ab$ **27** 3000
28 2100 **29** 880
30 252 **31** 15
32 150 **33** 264
34 68 **35** 38.4
36 630 **37** 140
38 (1) 20, (2) 60

Exercise 42 (p. 160)

1 $(x + c)(x^2 - cx + c^2)$ **2** $(y - a)(y^2 + ay + a^2)$
3 $(1 + 2a)(1 - 2a + 4a^2)$ **4** $(x - 4)(x^2 + 4x + 16)$
5 $(2 + 3c)(4 - 6c + 9c^2)$ **6** $(R - 1)(R^2 + R + 1)$
7 $(m - 5n)(m^2 + 5mn + 25n^2)$

8 $\left(xy + \dfrac{1}{2}\right)\left(x^2y^2 - \dfrac{xy}{2} + \dfrac{1}{4}\right)$

9 $\left(\dfrac{1}{x} + \dfrac{1}{y}\right)\left(\dfrac{1}{x^2} - \dfrac{1}{xy} + \dfrac{1}{y^2}\right)$

10 $\left(\dfrac{1}{x} - \dfrac{1}{y}\right)\left(\dfrac{1}{x^2} + \dfrac{1}{xy} + \dfrac{1}{y^2}\right)$

Exercise 43 (p. 163)

1 $\dfrac{y}{3xz}$

2 $\dfrac{3d^2}{4a^2}$

3 $\dfrac{x(x + y)}{2y(x - y)}$

4 $\dfrac{a^2 + 3b}{2ab(a - b)}$

5 $\dfrac{3}{2xy}$

6 $\dfrac{a + b}{a}$

7 $\dfrac{2}{x - 2}$

8 $\dfrac{x - 5}{x + 5}$

9 $\dfrac{a(a + 3b)}{b(a + 2b)}$

10 $\dfrac{x + y}{x^2 + xy + y^2}$

Exercise 44 (p. 164)

1 $\dfrac{x^2}{(x + 1)}$

2 $y(y + 1)$

3 $\dfrac{b}{a}$

4 $\dfrac{a - 7}{a - 3}$

5 $\dfrac{a(a + 2b)}{a - b}$

6 $\dfrac{(x + 1)(x - 3)}{x}$

7 1

8 $\dfrac{1}{2(x - 1)}$

9 $\dfrac{2(b - 4)}{b(b - 1)} ; \frac{1}{10}$

10 $\dfrac{x + 3}{2} ; 2.25$

Exercise 45 (p. 166)

1 $\dfrac{2x - y}{x(x - y)}$

2 $\dfrac{-(x + 1)}{(x - 1)(x - 2)}$

3 $\dfrac{13a - 5b}{6(a^2 - b^2)}$

4 $\dfrac{x^2 + 2x - 2}{x(x + 2)}$

5 $\dfrac{(3x3 + 7x^2 - 38x - 102)}{(x - 3)(x + 3)^2}$

6 $\dfrac{-5x^2 - 17x}{6(x + 1)(x - 2)}$

7 $\dfrac{7 - 3a}{(1 - a)^2}$

8 $\dfrac{12x}{x^2 - 9}$

9 $\dfrac{2y - x^2 - 2xy}{x(x + 2y)}$

10 $\dfrac{a(10b - 3a)}{3b(3a - b)}$

11 $\dfrac{t(a - b)}{(1 - at)(1 - bt)}$

12 $\dfrac{(x - a) + t(y - b)}{(a + bt)(x + yt)}$

13 $\dfrac{2x - 1}{x^2 - y^2}$

14 $\dfrac{-2y^2}{(x - y)^2(x + y)}$

15 $R = \dfrac{pq}{p + q}$

16 $R = \dfrac{pq}{q - p}$

17 $R = \dfrac{R_1 R_2}{2R_2 + 3R_1}$

18 $R = \dfrac{R_1 R_2}{3R_2 - 2R_1}$

19 $R = -\dfrac{r^2 - s^2}{2s}$

20 $R = \dfrac{5(p^2 - q^2)}{5p - q}$

21 $\dfrac{2PQ}{P + Q}$

22 $\dfrac{Q(P + Q)}{P - Q}$

23 1.47 to 2 decimal places

24 $P = \dfrac{2b + a}{2a + b} \cdot Q$

25 $n = \dfrac{IR}{E - Ir}$; 6

Exercise 46 (p. 168)

1	3	**2**	$\frac{5}{6}$	**3**	2.5
4	$-\frac{3}{5}$	**5**	-1.51786	**6**	$\frac{5}{3}$
7	7	**8**	0.3	**9**	4
10	6	**11**	$\frac{8}{5}$	**12**	6

13 8 **14** $V = \dfrac{RE}{R + r}$

Exercise 47 (p. 183)

Answers involving decimals are mostly approximate.

1 (1) 5.3 (2) ± 2.65 (3) ± 1.87
2 (1) ± 1.8 (2) ± 2.83 (3) ± 3.46
3 (1) ± 2.45 (2) ± 2.24 (3) ± 1.41
4 (1) ± 1.41 (2) ± 2.45
5 (1) 3.73 or 0.27 (2) -4 or -2
  (3) 4.24 or -0.24
6 (1) -3.41 or -0.59 (2) -3 or -1
  (3) 0.83 or -4.83
7 Min. value is -4 when $x = 3$.
  (1) 5 or 1 (2) 6.16 or -0.16
8 Min. value is -2 when $x = 2$.
  (1) 4.45 or -0.45 (2) 5 or -1
9 Min. value is -1.1 approx. when $x = \frac{5}{4}$. On substituting $x = \frac{5}{4}$ in the function, this equals $-1\frac{1}{8}$.
  (1) 2 or $\frac{1}{2}$ (2) 2.69 or -0.19
10 Max. value is 2.25 when $x = -0.5$.

Exercise 48 (p. 185)

1 (*a*) $x < 1$ or $x > 2$ (*b*) $1 < x < 2$
2 (*a*) $x \leqslant -3$ or $x \geqslant 1$ (*b*) $-3 < x < 1$
3 $x < -1$ or $x > 4$
4 $-5 \leqslant x \leqslant 1$

Exercise 49 (p. 187)

1 Check: $5 > 1 - 4 + 3 \; (= 0)$
2 Check: $5 > 1 - 4 - 5 \; (= -8)$
3 Check: $3 > 1^2$ and $3 < 2 \times 1 + 3$
4 (3, 1)

Exercise 50 (p. 192)

1	± 2	**2**	$\pm\frac{1}{2}$
3	± 12	**4**	$1, -3$
5	$8, -2$	**6**	$1, -11$
7	$\frac{3}{4}, -\frac{5}{4}$	**8**	16.75, 1.25
9	$8, 2$	**10**	$3, -4$
11	$5, -3$	**12**	$4, -7$
13	$8, -4$	**14**	$2, \frac{3}{2}$
15	$\frac{5}{2}, -1$	**16**	$3.25, -0.92$
17	1.434, 0.232	**18**	4 and -1.5
19	$5, -\frac{1}{3}$	**20**	$2.732, -0.732$
21	$13.14, -1.14$	**22**	$6.59, -7.59$

Exercise 51 (p. 195)

1	$0, 3$	**2**	$0, -5$	**3**	$2, 2$
4	$1, 2$	**5**	$1, -4$	**6**	$3, -7$
7	$\frac{7}{3}, -\frac{11}{2}$	**8**	$4, 5$	**9**	$2, -3$
10	$5, -7$	**11**	$-3, -10$	**12**	$11, -6$
13	$9, -5$	**14**	$\frac{3}{2}, 4$	**15**	$\frac{5}{2}, -1$
16	$1, \frac{1}{3}$	**17**	$\frac{5}{2}, -\frac{3}{4}$	**18**	$\frac{1}{4}, -\frac{2}{3}$
19	$2, 4, 5$	**20**	$\frac{1}{2}, -3, -2$	**21**	$1, 4, -2$
22	$\frac{5}{2}, 10, -5$	**23**	a, b	**24**	$\dfrac{c}{2}, -d$

Exercise 52 (p. 198)

1	$0.303, -3.303$	**2**	4.561, 0.438
3	$1.48, -1.08$	**4**	1.43, 0.23
5	$2.85, -0.35$	**6**	$1.744, -0.344$
7	$0.535, -6.535$	**8**	$3.441, -0.775$
9	$0.427, -2.927$	**10**	$0.803, -2.803$
11	$1.081, -1.481$	**12**	$-0.130, -0.770$
13	$2.591, -0.257$	**14**	$3.68, -0.43$
15	$9.75, -1.75$	**16**	$2.13, -5.62$
17	$6.123, -2.123$		

Exercise 53 (p. 200)

1	$\frac{5}{2}$ or $\frac{2}{5}$	**2**	9 mm, 15 mm
3	6.53 or -1.53	**4**	21.37, 27.37
5	5.54. Negative root has no meaning for the problem.		

6 79.78 or 0.209. One of the answers should be excluded on physical grounds.

7 12.5, -10

8 $-35/3$. This formula can represent the sum of an A.P. of n (integer > 0) terms (see section 202), so in that case $n = 35/3$ should be excluded.

9 15 mm, 20 mm **10** 144 m^2, 225 m^2

11 7, -1

12 12. Negative root is inadmissible.

13 108 m^2 **14** 100 m

15 8 and 12, or -12 and -8 **16** 12

Exercise 54 (p. 203)

1 $x = 6, y = 4; x = -5, y = -7$

2 $x = 5, y = 2; x = -26, y = 33$

3 $x = 2, y = 1; x = -\frac{7}{11}, y = \frac{69}{11}$

4 $x = \frac{1}{2}, y = 6\frac{1}{2}; x = 3\frac{1}{2}, y = -2\frac{1}{2}$

5 $x = -3, y = 2; x = -2, y = 1$

6 $x = 1, y = 4; x = \frac{4}{3}, y = \frac{34}{9}$

7 $x = 4, y = 3; x = -\frac{15}{14}, y = -\frac{171}{14}$

8 $x = -14, y = 11; x = \frac{9}{2}, y = -\frac{4}{3}$

9 $x = 7, y = 4; x = \frac{4}{3}, y = 21$

10 $x = 6, y = 4; x = -\frac{6}{13}, y = -\frac{116}{13}$

11 $x = 2, y = 4; x = \frac{4}{9}, y = -\frac{2}{3}$

Exercise 55 (p. 206)

1 $x = 3, y = 5; x = 5, y = 3$

2 $x = 7, y = 2; x = 2, y = 7$

3 $x = 8, y = 3; x = -3, y = -8$

4 $x = 1, y = 4$ or $x = 4, y = 1$

5 $x = 6, y = 2; x = -4, y = -3$

6 $x = -4, y = -1$ or $x = -1, y = -4$ or $x = 1, y = 4$ or $x = 4, y = 1$

7 $x = 9, y = 2; x = 2, y = 9; x = -2, y = -9; x = -9, y = -2$

8 $x = 7, y = 1; x = 1, y = 7$

9 $x = 6, y = 3; x = -6, y = -3; x = 3, y = 6; x = -3, y = -6$

10 $x = 3, y = 1; x = 1.5, y = 2$

Exercise 56 (p. 208)

1 (a) $x < 2$ or $x > 3$ (b) $-2 < x < 4$

(c) $x \geqslant -2$ or $x \leqslant -3$

2 (a) $x < 1$ or $x > 2$

(b) $x < 1$ or $x > 5$ (c) $x \le -1$ or $x \ge 4$

3 (a) $1 < x < 3$ (b) $-2 < x < 0.25$

   (c) $x \le -0.5$ or $x \ge 0.33 \ldots (1 \div 3)$

4 (a) $x > 3$ or $x < 2$ (b) $x < -0.5$ or $0.25 < x < 3$

5 (a) $0 < x < 3$ (b) $0 \le x \le 5$

   (c) $x > 5$

Exercise 57 (p. 211)

1	$6a^9$	**2**	a^{12}	**3**	$\frac{1}{8}a^8$
4	$2^{10} = 1024$	**5**	$4^6 = 4096$	**6**	$6a^8b^7$
7	$x^6 y^{12} z^4$	**8**	x^{2m}	**9**	a^{2m}
10	$a^{2p+q}b^p$	**11**	a^{2n}	**12**	a^{m+n+4}
13	a^3	**14**	$\frac{1}{2}x^3$	**15**	$3a^5$
16	$2^6 = 64$	**17**	$3^3 = 27$	**18**	x^5
19	$-5a^3b$	**20**	a^p	**21**	$4x^p y^q$
22	a^7	**23**	x	**24**	a^{n+1}
25	a^{2n}	**26**	a^6	**27**	$\dfrac{8ab}{9}$
28	$\dfrac{a^2}{b^2}$	**29**	$2^6 = 64$	**30**	$3^6 = 729$
31	x^{10}	**32**	x^{10}	**33**	a^{16}
34	$4a^4 b^6$	**35**	$\dfrac{x^{12} y^9}{8}$	**36**	a^{3p}
37	x^{4n}	**38**	$27a^{6p}$	**39**	$\pm a^2$
40	$\pm x^4$	**41**	$\pm x^8$	**42**	$\pm 3a^3$
43	$\pm a^2 b$	**44**	$\pm \dfrac{x^3}{y^2}$	**45**	$\pm \dfrac{3a^4}{2b^3}$
46	$\pm x^n$				

Exercise 58 (p. 218)

1 (1) $\sqrt[4]{3}$ (2) $\sqrt[5]{3}$ (3) $\sqrt[3]{a}$

   (4) $\pm\sqrt{2}$ (5) $\pm\sqrt{3}$ (6) $\sqrt[4]{a^3}$

2 (1) $\sqrt[4]{1000}$ (2) $\sqrt[3]{16}$ (3) $\sqrt[4]{a^5}$

   (4) $10\sqrt{10}$ (5) $4\sqrt{2}$ (6) $10\sqrt[4]{10}$

3 (1) $2\sqrt[5]{2}$ (2) $4\sqrt[4]{2}$ (3) $\sqrt[4]{a^7}$

   (4) $\sqrt{a^7} = a^2\sqrt[3]{a}$ (5) $\sqrt[5]{a^6} = a\sqrt[5]{a}$ (6) $\sqrt[5]{a}$

4 (1) 8   (2) 32   (3) 27
    (4) 1000   (5) 31.6   (6) $\pm\frac{1}{2}$

5 $\frac{1}{9}, \frac{1}{3}, 1, 3, 3\sqrt{3}, 9, 9\sqrt{3}$

6 (1) 5.656   (2) 27   (3) $\frac{1}{10}$
    (4) $a^{\frac{17}{12}}$ (5) 2   (6) $a^{1.2}$

7 $\frac{1}{4}, \dfrac{3}{a^2}, \dfrac{1}{\sqrt{2}}, 3, 3a^2, \dfrac{1}{1000}$

8 (1) $\dfrac{1}{x^3}$ (2) $\dfrac{1}{x^{\frac{1}{2}}}$ (3) x^3

    (4) $\dfrac{1}{x^6}$ (5) x (6) $\frac{1}{2}x^{\frac{1}{2}}$

9 (1) 4   (2) 125   (3) 1000

    (4) $\dfrac{1}{5^6} = \dfrac{1}{156\,25}$ (5) 16   (6) 3

10 (1) 4   (2) $\frac{27}{8}$ (3) $\pm\frac{1}{4}$
    (4) $\pm\frac{1}{6}$ (5) 8   (6) $\frac{1}{32}$

11 5.656

12 (1) $\sqrt[6]{a^5}$ (2) $100\sqrt{10} = 316.2$ (3) $\dfrac{1}{a^3}$

    (4) $\dfrac{1}{a^2}$ (5) x^{3n} (6) x^{2n}

13 (1) a^3 (2) 1   (3) 0

14 (1) 3.52×10^9 (2) 1.633×10^7 (3) 2.2×10^2
    (4) 3.7×10^3 (5) 2.52×10^{-7}

Exercise 59 (p. 226)

1 (a) 3; (b) 3   **2** (1) 1.5261   (2) -2.7806

3 1200   **4** 0.8268

Exercise 60 (p. 232)

1 (a) $100a : b$ (b) $\dfrac{3600p}{60q + r}$ **2** $3.6 \times 10^6 : 1$

3 (1) $\frac{8}{5}$ (2) $25 : 64$ (3) $5 : 12$

4 (*a*) 35   (*b*) $\dfrac{bx}{a}$ (*c*) $\dfrac{a}{x}$

5 $\dfrac{c(b+c)}{a}$ **6** $\dfrac{pA}{p+q}$ mm^2, $\dfrac{qA}{p+q}$ mm^2

7 $\dfrac{ax}{x+y}$ kg, $\dfrac{ay}{x+y}$ kg   **8** $\dfrac{bc}{a}$ or $\dfrac{c^2}{b}$ **9** 17

10 (1) $b\sqrt{ac}$ (2) $4ab$ (3) $(a+b)\sqrt{ab}$

11 $\frac{7}{3}$ **12** (1) $\frac{7}{3}$ (2) $\frac{3}{7}$ (3) $\pm\frac{5}{4}$

13 (*a*) 450   (*b*) $\dfrac{6a}{5b}$ (*c*) $\dfrac{4a}{3b}$

14 $x = \sqrt[3]{a^2b}$, $y = \sqrt[3]{ab^2}$

Exercise 61 (p. 240)

1 (*a*) No. The runner's rate differs at various parts of the race.
  (*b*) Yes   (*c*) No   (*d*) Yes
  (*e*) No. The connecting law is as explained in Section 178.
2 $\frac{8}{7}$; $45\frac{5}{7}$ **3** 37.5; 3.36
4 $y = \frac{33}{7}x$; $4\frac{5}{11}$ **5** $s = 3.2t$; 8.96 m
6 $E = \frac{8}{15}W$; 3.73   **7** $y = \frac{1}{2}x + 6$
8 $E = 0.06W + 0.2$ **9** $F = 0.42W - 0.6$

Exercise 62 (p. 248)

1 $y = \frac{8}{3}x^2$; $64\frac{2}{3}$ **2** $y = \frac{1}{4}x^3$; $\frac{27}{4}$
3 y values 12, 3; x value 6   **4** $y = \frac{7}{4}\sqrt{x}$; 8.75
5 384; $2\sqrt[3]{4}$ **6** $y = \dfrac{30}{x}$; 1.5

7 x values 2, 15; y value 10   **8** $F = \dfrac{1920}{d^2}$

9 4 s   **10** 10 m
11 102 mm   **12** 1.8%
13 328 s   **14** $y = 8x^{1.4}$

Exercise 63 (p. 252)

1 (a) $y \propto xz$ $\therefore y = kxz$

  (b) $y \propto \dfrac{x}{z^2}$ $\therefore y = \dfrac{kx}{z^2}$

  (c) $y \propto \dfrac{\sqrt{x}}{z}$ $\therefore y = \dfrac{k\sqrt{x}}{z}$

  (d) $V \propto hr^2$ $\therefore y = khr^2$

  (e) $W \propto \dfrac{bd^2}{l}$ $\therefore W = k \cdot \dfrac{bd^2}{l}$

  (f) $y \propto \dfrac{x^2}{\sqrt[3]{z}}$ $\therefore y = k \cdot \dfrac{x^2}{\sqrt[3]{z}}$

2 $y = \dfrac{25}{4}\,\dfrac{x}{z}$, 15   **3** $y = 3xz^2$; 8

4 $10\frac{15}{4}$ **5** 8 t   **6** 1

7 $H = \dfrac{0.24tE^2}{R}$ (1) 24 000 units   (2) 150 s

8 4.1% increase.

Exercise 64 (p. 259)

1 $y = 6x^2 + 3$ **2** $y = 0.42x^2 + 7.5$
3 $R = 0.01V^2 + 5$ **4** $H = 1.07V^3 + 441.1$

5 $y = 0.54 + 0.24x^2$ **6** $R = \dfrac{1200}{V} + 50$

7 $y = 0.25x^3$ **8** $y = 3x^{\frac{1}{3}}$
9 $Q = 5.7H^{0.36}$ **10** $y = 3.665x^{0.6}$

Exercise 65 (p. 265)

1 (1) $\sqrt{150}$ (2) $\sqrt{288}$ (3) $\sqrt{500}$
   (4) $\sqrt{208}$ (5) $\sqrt{4a^2b}$ (6) $\sqrt{9x^4y}$
2 (1) $20\sqrt{2}$ (2) $8\sqrt{5}$ (3) $6\sqrt{3}$
   (4) $50\sqrt{2}$ (5) $5\sqrt{15}$ (6) $60\sqrt{2}$
   (7) $2ab\sqrt{6a}$ (8) $3x^2y\sqrt{2yz}$ (9) $5abc\sqrt{3abc}$

(10) $10p\sqrt{10q}$ (11) $a(a-b)\sqrt{b(a-b)}$

(12) $5y(x-2y)\sqrt{xy}$

3 (1) $3\sqrt{5}$ (2) $7\sqrt{2}$ (3) $16\sqrt{3}$

   (4) $10\sqrt{10}$ (5) $4-\sqrt{2}$ (6) $7\sqrt{2}+\sqrt{14}$

   (7) $(a+b)\sqrt{a-b}$ (8) $3(x-2y)\sqrt{2(x+2y)}$

4 (1) $2\sqrt{2}-1$ (2) $2-3\sqrt{6}$ (3) $3-2\sqrt{2}$

   (4) $23+4\sqrt{15}$ (5) $57-10\sqrt{14}$ (6) 2

   (7) 38 (8) $301+20\sqrt{3}$ (9) $a-25$

   (10) 21

5 (1) $\dfrac{2\sqrt{3}}{3}$ (2) $\dfrac{\sqrt{5}}{5}$ (3) $\dfrac{3\sqrt{7}}{14}$

   (4) $\dfrac{\sqrt{10}}{30}$ (5) $\dfrac{\sqrt{2}}{8}$ (6) $\dfrac{6\sqrt{5}}{5}$

   (7) $\dfrac{\sqrt{6}}{3}$ (8) $\dfrac{2\sqrt{35}}{7}$ (9) $\dfrac{\sqrt{5}+1}{4}$

(10) $\dfrac{3(\sqrt{7}-\sqrt{2})}{5}$ (11) $\dfrac{3(2\sqrt{3}+\sqrt{2})}{10}$

(12) $5(2\sqrt{7}+3\sqrt{3})$ (13) $5+2\sqrt{6}$

(14) $6-\sqrt{35}$ (15) $\dfrac{21+\sqrt{7}}{62}$

(16) $2\sqrt{2}$ (17) $-\sqrt{3}$

(18) $\dfrac{4(6+\sqrt{5})}{31}$

Exercise 66 (p. 271)

1 (1) $12.5, 15, 17.5$ (2) $0, -4, -8$

   (3) $(a-3b), (a-5b), (a-7b)$

   (4) $6.6, 7.9, 9.2$

   (5) $(x+2y), (x+3y), (x+4y)$

2 $12, 16.5$ **3** $4+4p$

4 $x+3n-1$ **5** 3.48

6 29 **7** 2 **8** 19

9 (*a*) 217.5 (*b*) 16 (*c*) 0 (*d*) $364\frac{1}{2}$

10 12 **11** 5 or 8 **12** $120; 1860$

13 £400.50; £53,437.50   **14** £3400

Exercise 67 (p. 275)

1. (a) 62.5, 156.25, 390.625   (b) $\frac{1}{4}, \frac{1}{16}, \frac{1}{64}$
   (c) $-54, 81, \frac{24\frac{3}{4}}{2}$
   (d) $3 \times 10^{-4}, 3 \times 10^{-5}, 3 \times 10^{-6}$
   (e) 0.010 125, 0.001 518 75, 0.000 227 8 (approx.)
2. 320   3. $\frac{128}{234}$
4. 1.610 51   5. 0.001 215
6. (a) $ar^{2n-1}, ar^{2n}, -ar^{2n-1}, ar^{2n}$
7. 1.05   8. ± 1.04
9. (1) 3.873; (2) 3.888 (both approx.)
10. 7, 9.8   11. £12,155.10   12. £162 901.25
13. 1.04   14. 10, 20, 40

Exercise 68 (p. 277)

1. (a) 94.5   (b) 19.92   (c) $\frac{21}{64}$
2. $\frac{315}{32}$ 3. $\frac{133}{243}$ 4. 16.32
5. 216.8   6. 765   7. $27, 40\frac{40}{81}$
8. 158.702

Exercise 69 (p. 283)

1. (a) $\frac{2}{3}$ (b) $\frac{1}{3}$ (c) 1
2. (a) $\frac{10}{99}$ (b) $\frac{6}{99}$ (c) $\frac{16}{99}$
   Series (c) is the sum of (a) and (b). It represents 0.161 616..., a recurring decimal.
3. $\frac{2}{3}$. This series equals the series (b) in question (1) increased by unity and then diminished by series (a).
4. In $\frac{3}{2}\{1 - (\frac{1}{3})^n\}$, $(\frac{1}{3})^n \to 0$ as $n \to \infty$. $\therefore$ the limit is $\frac{3}{2}$.
5. $\dfrac{b}{b - a}\left\{1 - \left(\dfrac{a}{b}\right)^n\right\}$; limit $= \dfrac{b}{b - a}$
6. (a) $\frac{16}{5}$ (b) $\frac{25}{6}$ (c) $\frac{27}{5}$
7. (a) 26   (b) $2(\sqrt{2} + 1)$
8. $\frac{4}{5}$ 9. 190 m   10. £3.3 $\times 10^7$ to 2 significant figures

Exercise 70 (p. 288)

Most of the answers are approximate.

1. £11,800   2. £55,000   3. £4,860
4. £24,300   5. £34,370   6. £50,000
7. £166,700   8. £240,000

Index